Skype: Bodies, Screens, Space

Despite the popularity of Skype with video many of us are still figuring out how to 'do' it. Interviews reveal that we can now run the programme but we are less certain about how to 'perform' in front of the webcam. Seeing ourselves in the box on the side can feel strange. We are not quite sure which bits of our bodies to display on the screen, how much to move around the room, or move the device around the room. Is it acceptable to use Skype with video at a funeral, in crowded spaces or while in bed? This book addresses how people are emotionally and affectually connecting with others audio-synchronously on the screen in a variety of different spatial contexts. Topics include Skype with video being used by grandparents to connect with grandchildren, friends and family using it for special occasions, and partners using it for romance and sex. Theories addressing bodies, gender, queerness, phenomenology and orientation inform the research. It concludes that while Skype does not offer some kind of utopian future, it does open up possibilities for existing power relations to be filtered through new lines of sight/site which are shaping what bodies can do and where.

Robyn Longhurst is Deputy Vice-Chancellor, Academic and Professor of Geography at University of Waikato. She has served as Editor-in-Chief of *Gender, Place and Culture: A Journal of Feminist Geography* and Chair of the International Geographical Union Commission on Gender and Geography. Robyn has published on issues relating to digital media, pregnancy, mothering, sexuality, 'visceral geographies', masculinities, and body size and shape.

Skype: Bodies, Screens, Space

Robyn Longhurst

Routledge
Taylor & Francis Group

LONDON AND NEW YORK

First published 2017
by Routledge
2 Park Square, Milton Park, Abingdon, Oxon OX14 4RN

and by Routledge
711 Third Avenue, New York, NY 10017

First issued in paperback 2018

Routledge is an imprint of the Taylor & Francis Group, an informa business

British Library Cataloguing in Publication Data
A catalogue record for this book is available from the British Library

Library of Congress Cataloging in Publication Data
Names: Longhurst, Robyn, 1962-
Title: Skype : bodies, screens, space / Robyn Longhurst.
Description: Abingdon, Oxon ; New York, NY : Routledge, 2017. | Includes
bibliographical references and index.
Identifiers: LCCN 2016016607 | ISBN 9781472434548 (hardback) |
ISBN 9781315609294 (ebook)
Subjects: LCSH: Skype (Electronic resource)
Classification: LCC TK5105.8865 .L66 2017 | DDC 384.3/8–dc23
LC record available at https://lccn.loc.gov/2016016607

ISBN 13: 978-1-138-60110-9 (pbk)
ISBN 13: 978-1-4724-3454-8 (hbk)

Typeset in Times New Roman
by Taylor & Francis Books

Contents

Illustrations

Figures

Table

Preface

The book can be read from beginning to end, from front to back, but I am aware that many of us in our busy daily lives dip in and out of books, or more commonly in and out of websites, e-mail, Facebook, Instagram and Twitter. I have written the book with this in mind. The chapters are not wholly self-contained but readers with an interest in, for example, sex and sexuality may be interested in Chapter 7 on 'Skype sex' whereas someone with an interest in intergenerational relationships might be more interested in reading Chapter 5 on 'Family, friends and loved ones' but not dip into the rest. Some themes, however, thread through all of the chapters, for example, that queer phenomenology can enable us to think more deeply about how Skype might be providing new angles on subjectivities and relationships. In some instances this can feel disorientating. When technologies are new it is difficult to 'sink' into them. They do not fade into the background. We tend to notice them and respond in ways that at times seem awkward. This is discussed in detail in Chapter 2 'Queer phenomenology', referred to throughout other chapters, and returned to in more detail in the final chapter, Chapter 8 'Reorientating bodies and spaces'.

Other pieces do not fit together as readily. The book is not a carefully crafted overview of people's use of Skype, with one chapter leading directly to the next. In some ways it is more like Daniel Miller's (2011) book *Tales from Facebook* which provides readers with a series of 'tales' about Facebook illustrating how it can become the means through which people can nurture relationships, but also break up marriages, isolate people in their homes and so on. A few years after the publication of Miller's book on Facebook, Miller and Sinanan (2014) published *Webcam*. This book builds on *Webcam* in an attempt to dig a little deeper to demonstrate how the specific digital media of Skype with video is not simply transforming lives but is making bodies, spaces and screens seem, for now at least, a little queer or 'strange'.

Finally, one more thing that will likely become apparent to readers as they engage with this text is that it draws from a range of disciplines. Indeed, one of the pleasures but also frustrations of working in the area of digital media is that it is highly interdisciplinary. I certainly do not profess to be an expert across all the fields, for example, human geography, feminist studies, queer studies, digital anthropology, and screen and media studies which intersect around this area. This book, therefore does not offer readers a comprehensive reading of Skype from one or even two disciplinary areas (although human geography does provide something of a focus) but instead it tends to skirt over many areas and instead, largely through Sara Ahmed's queer phenomenology and the stories of interviewees, offers some insights which will enable readers to catch glimpses of how Skype might be 'queering' bodies, screens and space.

Acknowledgements

First and foremost, I want to thank the research participants who contributed their stories of using Skype for inclusion in this book. As many of you know, stories can be profound and are a powerful mechanism through which we learn. The ones gifted to me for this project were no exception. I realize now, in retrospect, that I began this project with a fairly narrow idea about how people use Skype. After hearing many of the stories I was forced to rethink things, which is of course the point of research. Not only did research participants tell me their stories but so too did others, for example, friends and strangers at social functions. People often asked me: 'So, what are you are researching?' When I replied 'Skype' people would kindly share their stories, providing a rich context for this project.

I also want to thank colleagues for the various roles they played in this book's production. It is only fitting that I begin with Professor Lynda Johnston whose personal, collegial and intellectual support over more than two decades has been immeasurable. I am also indebted to Lynda for her astute commentary on a draft of this book. There has never been a moment when I have not thoroughly enjoyed working with Lynda.

There are many other colleagues to whom I am also grateful and who have provided invaluable assistance, albeit at different times, and in different ways. In particular, I would like to thank Associate Professor John Campbell for many years of conversation in the Waikato 'Geography Tea Room' where no subject is off limits. The many fantastic women colleagues in my programme at Waikato also deserve special thanks, including Dr Naomi Simmonds and Dr Gail Adams-Hutcheson, both of whom were my PhD students. They now teach and research at Waikato, which is wonderful for many reasons, not least that it means I am able to enjoy their company as colleagues. I am also grateful to Dr Elaine Bliss, Associate Professor Lex Chalmers, Dr Anne-Marie d'Hauteserre, Betty-Ann Kamp, Dr Colin McLeay, Max Oulton and Dr Pip Wallace for all the intellectual and warm conversation over a period of many years.

Students in the graduate geography course 'Crossing Boundaries' at the University of Waikato also deserve thanks for engaging in much thought-provoking discussion about 'bodies, screens and space'. Engaging with the many wonderful students who have taken this course over more than a decade that Professor Lynda Johnston and I have taught it has been one of the highlights of my academic career.

A number of research assistants, including Cherie Todd and Rebecca Campbell, who played a vital role in collecting interview data and Sunita Basnet, who searched data bases, pulled together information and edited the list of references, also deserve my special appreciation. Each one of you has done a wonderful job and I am most grateful.

I began writing this book at the same time as I took on a new role in the senior leadership team at the University of Waikato. In retrospect this was probably ill advised given the time commitments involved in such roles! Colleagues, however, supported me and kept me smiling as I juggled far too many meetings, a pile of administration, numerous events in the evening, and of course writing. Professor Neil Quigley, Professor Alister Jones, Professor Bruce Clarkson, Duanna Fowler and Joseph McFarlane with whom I work closely on a daily basis deserve special mention. Brenda Hall, Paula Maynard, and more recently Carole Luckwell, Keely Smith, Marie Ward and Tamara Deverson have offered me excellent administrative support during the time of writing this book enabling me to focus on the job at hand.

I would also like to acknowledge some people whom I have never met but whose ideas have profoundly influenced me, although I hasten to exempt them from any responsibility for how I have used their ideas. In particular, I owe a debt to Professor Sara Ahmed whose ideas I have drawn on heavily in this book. I have never had the pleasure of meeting Professor Ahmed but I have had the pleasure of engaging with her research, which is both inspiring and empowering. The work of other feminist, queer, and critical scholars such as Elizabeth Grosz, Judith Butler, Gillian Rose, Gill Valentine and Linda McDowell has also helped shape my thinking on gender, sex, space and place over a period of many years. I am also grateful to scholars who have made a profound contribution towards my thinking on digital media, namely Professor Daniel Miller and Dr Jolynna Sinanan, whose research on webcam I draw on extensively in this book.

I would also like to thank two editors at Ashgate, Valerie Rose, who I contacted in the first instance about my idea for the book, followed by Commissioning Editor Katy Crossan. Both were from the outset enthusiastic about and supportive of this project. Towards the end of the writing Ashgate was acquired by Routledge's parent company, Informa. At this point I began working with Faye Leerink, Geography and Tourism Editor and Priscilla Corbett, Editorial Assistant, who I also wish to thank.

Material in Chapter 4 on using Skype to conduct interviews was presented by Dr Gail Adams-Hutcheson at the 5th International and Interdisciplinary Conference on Emotional Geographies in Edinburgh, Scotland, 10–12 June 2015 and at Sixth New Zealand Mobilities Symposium: Mobilities in a 'Dangerous World' at the University of Waikato in Hamilton, New Zealand, 25–26 June 2015. Some of the information in the first half of Chapter 5 on

Skyping across generations was presented in a keynote address titled 'Using Skype for intergenerational maternal encounters: bodies, spaces and screens' in a workshop 'Embodying the lifecourse' at University of Durham, Durham, UK, 21 November 2012. I am grateful for the generous discussion that followed this presentation. Some of the ideas contained in Chapter 7 were presented in a paper titled 'Skype sex, love and romance' at the Association of American Geographers Annual Meeting in Chicago, USA, 21–25 April 2015. Again, I would like to thank members of the audience for their questions following the presentation.

Last but certainly not least, I offer my gratitude to my family – my mother Colleen for her enduring love, my sons Benjamin and Jerome for making me laugh and keeping me 'up with the play' with social media and popular culture, and my partner David for being my biggest supporter and preparing fantastic food when, left to my own devices, I would have simply defaulted to toast and vegemite!

1 Why Skype, why now?

It's 8 pm and I'm in a hotel room in Sydney, Australia having spent the day at a seminar on 'blended learning' (combining classroom and online learning). It's long past dinner time but I've not yet managed to eat so have ordered room service – burger and fries. Hunger has led to calories trumping nutrition. The meal arrives just as I am Skyping my 18 year old son in Auckland, Aotearoa New Zealand.[1] He recently moved out of home, into a university self-catering halls of residence. The middle-aged hotel worker sets the meal down on the desk beside my laptop. I comment: 'I'm just talking to my son on Skype.' My son calls out 'Hi' to the man delivering the meal. He doesn't respond, seeming uncertain about the protocols surrounding greeting a stranger in a hotel room who exists only as an audio-visual presence on a laptop screen. After the hotel worker leaves the room I sit back down at the screen to pick up where I left off with my son. At the same time I crunch on a fry. 'That's just cruel' he quips. 'I might not be able to smell your dinner but I can see and hear you.' The hungry teenager in New Zealand is still learning to cook and budget constraints mean there is never very much in the pantry. The burger and fries is twice as much food as I can eat but to state the obvious, sharing via Skype is not an option (Author's research notes, 25 April 2014).

Last night approximately 35 people gathered in a large staff tea-room at my University for a book launch. Seats were arranged in a series of semi-circles, beverages and snacks were available and at the front of the room was a large screen on a portable trolley. Present on the screen from Indiana, USA via Skype was a professor who was a co-author of one of three books being launched that evening. The other co-author was physically present in the room. The image of the on-screen author, a middle-aged man wearing glasses and a purple suit jacket, was crystal clear. There were no technical glitches or problems with placing the call, delays in audio or visual transmission, stilted or pixelated images, all of which I have grown accustomed to at events such as this where Skype is used. Tonight was different. In New Zealand it was early evening. In Indiana it was early hours of the morning. The professor on screen, however, did not appear tired, rather he was enjoying a glass of wine with the rest of us, displaying at one point to the camera the bottle he'd poured the wine from. It quickly became apparent that the author on screen was very comfortable with the technological interface. At the book launch the professor on Skype was given an opportunity to 'hold the floor' (via the screen) both with a speech about his new book and interjecting when appropriate often with humorous one-liners during others' presentations. To help him engage the audience from afar the professor had, hidden from view until he required them, a number of props. He seemed to have a sense that being on a screen in a large room full of real people meant that it would be easy for him to become invisible. So, he showed us all a pink suit which he'd considered wearing (instead of the purple one). He changed into dark glasses at a couple of junctures during the evening and showed the audience a globe pointing to Fiji to highlight his narrative about the new book's global circulation. When another 'real' speaker mentioned business ethics he pulled out a handful of fake US dollars displaying them to camera. The audience laughed at his numerous interjections. The professor felt present in the room, it felt like he and others were orientated in a way that was comfortable. This is so often not the case when people Skype in for events such as this (Author's research notes, 1 August 2014).

These two stories highlight a number of points that are pertinent to this book. In the first story on the day in question I had spent many long hours discussing the merits of 'blended learning'. There seemed to be general consensus amongst our group of approximately 40 that blending online with traditional face-to-face class activities in a planned, pedagogically valuable way has a great deal to offer learners. Over the past few years educators have discussed at length the potential pros and cons of using digital media, including Skype, especially with younger learners, to not just supplement but transform the learning process. Much less has been written about how using digital media might be transforming not just teaching and learning but also everyday interactions between bodies,[2] time and space are being reorientated. In Aotearoa New Zealand where I live and where this research took place, more than three quarters of the population use the Internet (Bell et al. 2008) including to make Skype calls with video. Having real-time audio and visual contact with my son but not being able to touch him, to share food or even the smell of food with him prompts a particular embodied experience that many of us who use Skype are still trying to fully comprehend.

In the second story I had not been expecting to be so thoroughly engaged by the professor on the screen. In fact, although I had used Skype with video many times before both in a personal and professional capacity I had never thought to use it in this way – exaggerating expressions and producing props – to engage with others. The professor used his body, consciously and carefully orientating it towards the camera in ways that I had not been expecting and as a result he made the event 'work'. Up until that point many of my experiences of using Skype at work to connect with colleagues, job applicants and students could be described at worst as failures (technology did not work, one or both parties were disengaged, the link dropped) and at best as okay (we managed to connect although the sound and picture quality were poor). This time was different and made me think about bodies on-screen and off-screen, their orientation to the camera and to the spaces in which they reside.

I start with these personal stories because the process of constructing knowledge is always embodied (Butler 1993; Grosz 1994). Readers will note that throughout the book I draw on my own experiences, personally and professionally. I connect back to research I've conducted in the past. This is in an attempt to breathe life into the theoretical points being made. Feminists have long argued that our subjectivities are made and remade through our research (Rose 1997) and that the personal and professional, past and present, private and public are not easy to separate. As I sat at the kitchen table typing this manuscript I had Skype turned on and occasionally received alerts when particular family members or colleagues were online. At times I simultaneously wrote about Skype on one screen while I talked with someone using Skype visible on another screen. Research and researchers are not easily separable. This does not mean that we can all assume to somehow easily know and represent ourselves but reflecting on our own complex and shifting embodied practices can help materialize the notion that our subjectivities are gendered, sexed, raced, that we are a particular body size and shape and so on both online and offline.

I am sure that it will not come as news to anyone who reads this book that this materiality of our lives and bodies over the past few years has come to be increasingly filtered through screens (e.g. in relation to gender see Fluri 2006; Fortunati and Taipale 2012; Friedman and Calixte 2009; Frizzo-Barker and Chow-White 2012; Wajcman 2004, 2007). Skype does not have inherent capacities or qualities as a digital media that prompts particular interactions but it does orientate us in particular directions, it mediates our corporeality, emotions and affects (Davidson and Bondi 2004; Davidson and Milligan 2004; Thien 2005, 2011, Thrift 2004) and therefore makes a difference to our interactions in the same way that offline space makes a difference to our interactions. How exactly, and in what ways, though remains uncertain.

There are many examples on the internet including in blogs, YouTube, Twitter and Instagram of the many ways in which people are using Skype (e.g. see #SkypeSavvy set up by Skype in 2015 for people to share the numerous ways in which they are using Skype) but still little is known about some of the orientations, feelings, emotions and affects which accompany this use (Davidson, Bondi and Smith 2005) including the ways in which they are embedded in power relations or power 'geometries' (Tolia-Kelly 2006). This book aims to make some headway on these issues. I am interested in how bodies, through Skype with video, are orientated in space and how their materiality is filtered through the screen in real time.

I chose to focus on Skype because of its pervasiveness in the digital market and because it enables people to not just hear but *see* each other. I am aware that Skype is certainly not the only programme available. There are currently a range of other synchronous audio-visual software technologies which people are also currently using and which are mentioned as part of this research.

Some examples include Facetime, Scopia, Google Hangouts and Snapchat. I am interested in this range of audio-visual conferencing opportunities and so therefore have not excluded these from discussion but at the time of writing this book Skype was the preferred platform for video chat in Aotearoa New Zealand and in many other places. Indeed it is telling that Microsoft Skype has dominated the video chat world for so long that the name 'Skype' has morphed into a verb used in sentences such as 'When do you want to Skype?' and 'Will we be Skyping when you are overseas?'

Despite the popularity of Skype many of us are still learning how to 'do' it. I do not mean simply how to run the programme (although there is that too) but how to 'perform' (Butler 1990) in front of the webcam, how to emotionally and affectually connect with people on the screen, which bits of our bodies to display, highlight or hide on the screen, how much we can move around the room, or move the device around the room or further afield, whether it is 'appropriate' to live-stream Skype at a funeral or whether we ought to communicate audio-visually on our mobile devices in crowded trains, when in bed and so on. One of the participants in this study, who is aged in his early twenties, commented to me that if someone is 'old' Skype might be 'a bit weird' but if they have grown up with the internet being around them all the time then 'it should be pretty natural'. The participant makes a useful point about different generations' engagement with digital media but this research indicates that much more is going on in relation to Skype than can be explained by a generation gap.

This book examines bodies, screens and space in relation to Skype. It addresses questions such as, is space acquiring a new shape and feel through Skype? Are the contours of the spaces we know and inhabit changing as we spend more time interacting audio-visually and synchronously with family, friends, colleagues and lovers? Is Skype reorientating bodies in new directions? At times we may, for example, feel orientated and comforted by being able to see our loved ones on the screen but at other times we may feel disorientated and uncomfortable about not being able to reach out and touch them or smell the familiar scent of their body. How are the interactions through the medium of Skype – through screens – framing our bodies and spaces differently? Is it reorientating them in ways that most of us are not yet used to and if so what might be some of the implications of this? These are all questions that it seems timely to pause and reflect upon.

Feeling my way

I begin this section with how Skype *feels*[3] – some of the emotions (Parr 2005) and affects (Anderson 2006) that surround Skype – because it is this that has driven so much of the enquiry over the past few years. In referring to emotions and affects I am not meaning to separate out emotion (as an expression of individual subjectivity) and affect (as a collective expression) but instead see them as intertwined (Ahmed 2004; Lalibertè and Schurr 2015; Sharp 2009;

Wright 2010). For me, Skype generates emotions and affects that at times feel very ordinary – mundane. Skyping is something that millions of people across the world do for a variety of reasons on a daily basis. At other times it feels utterly strange – queer – the person on the screen is both present and absent. I marvel at the technology that can connect me to others, both aurally and visually in real time. It might be with the neighbour next door or it might be with my sibling on the other side of the world.

Bruno Latour (1999) argues the more refined a technology becomes, the more it tends to disappear into the everyday, become unnoticeable. Skype at this particular historical juncture is not new but neither has it (yet) disappeared into the everyday becoming unnoticeable. To me, and I suspect to many of the participants in this research, it seems simultaneously both ordinary and extraordinary, old and new, orientating and disorientating. This makes it a good time to examine Skype.

Skype is currently just one of many media[4] which are reorientating bodies and spaces. Madianou and Miller (2012, 3) argue it is important to acknowledge 'that most people use a constellation of different media as an integrated environment in which each medium finds its niche in relation to others' and that we need to move attention away from the individual technical propensities of any particular medium to think instead about the way they work together as 'polymedia' (also see Nayar 2010). While this is a valid point Nancy Baym (2010, 17) notes that it is not necessary to always or only look at new media as a whole because 'Each of these media … offers unique affordances, or packages of potentials and constraints, for communication.' She continues that it is important to consider them both holistically and separately.

This book focuses mainly on one media, that is, on Skype, while acknowledging that in most cases it is used alongside and in conjunction with other media. Also, for the purposes of this research I am most interested in Skype's free service of calling *with video* (approximately half of all Skype calls employ video). Previous research I conducted on mothers' use of multiple media (Longhurst 2016) revealed that this was the form of communication that they were most excited about and keen to use.[5] It appears to be capturing people's (and not just mothers') interest hence my decision to pay it more detailed attention.

On their website Skype.com (2014, no page number) claim:

> Skype is for doing things together, whenever you're apart. Skype's text, voice and video make it simple to share experiences with the people that matter to you, wherever they are. With Skype, you can share a story, celebrate a birthday, learn a language, hold a meeting, work with colleagues – just about anything you need to do together every day. You can use Skype on whatever works best for you – on your phone or computer or a TV with Skype on it.

Skype promotes itself as a platform for 'doing things together' except this 'doing' is a new kind of 'doing' – doing that involves bodies sharing images, objects and experiences online. The promoters of Skype claim that it is 'simple to share experiences'. Participants in this study, however, report that sharing via Skype is often anything but simple. For some their stories sit in sharp relief to those posted on the Skype website, for example, by enthusiastic stay-at-home dad Jeff Bogle in a blog titled 'Time and space are no big deal, thanks to Skype'. Bogle (2014, no page number) writes that the only thing that has stopped people 'from being the loving son or daughter' who are available when a family member needs to cry on someone's shoulder is distance. It is not that people do not want to show love and affection but that physical distance has separated them. Bogle continues: 'We are fortunate to be alive in a modern age where time zones are nothing, technology is king, and what can happen in a movie, can in fact happen.' Like a number of other proponents of Skype, Bogle sees it as offering the future that many have long dreamed of. Skype opens up a new world that enables family members to chat over a cup of tea, share stories, laugh and feel connected even though they are separated by physical distance.

In fact, many bloggers, like Bogle, are quick to 'sing Skype's praises', for example, The Skype Team in Skype Moments of the Month note:

> By bringing people together, Skype is making the world a more open and cultured place. No matter if it's reuniting refugees with loved ones, sharing a new perspective at a film festival or a global collaboration between classrooms on different sides of the world, Skype acts as the uniting force (Skype.com 2015, no page number).

Again, this utopian reading of Skype simplifies things. Is the world really becoming 'more open and cultured' and if so, how exactly? Maybe Skype is uniting people but the emotions and affects produced by and through Skype are complex. As the participants' narratives will illustrate communicating through Skype does not necessarily simply erase physical distance and different time zones. It does not always unite people. It offers some a way of being together but this is a specific kind of being together, a specific kind of 'doing', a specific kind of communication which involves sharing not all, but some of the bodily senses. It is a new kind of communication, of 'intimate sensing' (Porteous 1986) that is governed by an emerging set of unwritten rules of social conduct and etiquette. Many of us are still learning 'Skype etiquette' – when and how to connect, whether to message someone first (a bit like knocking on the door before entering), how to show our bodies on the screen, which bits to conceal and which bits to highlight, whether we need to sit and stay still or move around the room, or go outside, whether there are certain spaces where it is not acceptable to Skype from and what these spaces might be. I am interested in understanding the multiple and intersecting ways that power operates through bodies (Foucault 1980; Sharp et al. 2000) and the digital media of Skype. The bodies and spaces engaged in Skype encounters appear both normative and non-normative. Space is embodied but 'flat', that is, somehow strangely disembodied. We see and hear but cannot touch and smell. Some bodily senses are present and privileged over others.

People's interactions over space and time become framed differently through Skype. It is not simply a question of the degree of framing. Skype, and other similar synchronous audio-visual communication technologies, are not just mediating more intensely existing social relations because all social relations are mediated (Horst and Miller 2012). Rather, many of us are having to rethink our social interactions. Even if we understand the mechanical process of setting up the programme and making the call we are not always comfortable with the array of interactions that then take place.

This is not surprising given that interactions (human and non-human) including those that take place via Skype involve a series of embodied performances (Butler 1990) which are structured by various dimensions of subjectivity including gender, sex, sexuality, age, ethnicity, culture, social class and (dis) ability. Two decades ago social research on the role of digital media tended to focus optimistically on the alleged potential of such media to address and alleviate inequalities. Now, however, more nuanced understandings of how technology and social structures shape each other are increasingly common. There have been calls for greater use of critical theory in understanding technology and digital media (e.g. Feenberg 1991, 1999; Fuchs 2009a, 2009b). It is now well established that the 'real'[6] and representation, the material and visual, offline and online space are inextricably linked (Kitchin 1998a, 1998b; also see Adams, 1997, 2007; Crang, Crang and May 1999). There is a simultaneity of offline and online worlds. Skype is not separate or disconnected from online space or the real world. It is evident that digital media such as Skype mean that people are continually reconfiguring their 'encounters with space, form, time, grammars of meaning and their habitual interpretation' (Rose and Tolia-Kelly 2012, 1–2). They are profoundly reshaping family lives, work lives and love lives. Millions use Skype on a daily basis to communicate, create and maintain relationships with partners, family members, friends and lovers. Skype has become a critical arena in which many people across a range of social classes, ethnicities, sexualities, genders, abilities and ages negotiate ideas and practices of daily life. And yet still relatively little is known of the ways in which Skype is refashioning imaginaries and practices of daily life, mediating the ways in which family relationships, collegial partnerships and 'intimacy' are experienced and lived.

It is important for researchers to find out more about the everyday uses of digital media which play a major role in so many people's lives. Some important work has been conducted (e.g. De Jong 2015 on storytelling and Facebook; Meunier 2010 on intimate online spaces) but more work remains to be done. Clearly Skype with video is just one way in which people are increasingly using technology to conduct social relations but it is for me one of the most interesting because unlike other technologies such as phoning (Horst and Miller 2006), emailing, Facebooking (Miller 2011), or texting (Thompson and Cupples 2008) it involves the transmission of a real-time visual image. Elisabeth Roberts (2012, 386), adopting what she describes as a 'hauntological approach', argues that 'visual images have an undecidable, "in-between" status, haunting between material and immaterial, real and virtual'. They are potentially powerful.

Many commentators who addressed the upsurge of interest in digital media more than a decade ago focused on the pitfalls and possibilities of reconfiguring identities and spaces when people are no longer tied to the specificity of their material body (e.g. see Hardey 2002 on 'embodiment and identity through the Internet') but Skype with video conveys something of the materiality of bodies. It is about hearing, and *seeing* the lived flesh of the real person or people on the screen. It is about being able to observe the expressions, comportment, clothing, movements and surrounds of others on screen. It is this that prompted me to focus on Skype.

As a geographer with a long-standing interest in mothering, 'the body' and space, I wanted to find out more about how people are using and feeling about this visual technology as a medium for conducting relationships – how it might be reorientating bodies and spaces. Knowledge of people's relationships with digital media is currently being generated from a variety of disciplines including digital humanities, cultural studies, media studies, anthropology, sociology, internet studies, feminist studies, queer studies and also human geography. In my case feminist studies, queer studies and human geography have enabled me to develop a spatial perspective that can cast light on orientations, objects, bodies, distances and lines.

In addition to theoretical work, I also draw on empirically-grounded work to flesh out understandings. I have enjoyed talking to people in order to find out more about how they are creating and maintaining social relationships through digital media, namely Skype. It is enabling many of us to expand our inter-personal communicative possibilities, to widen our social networks beyond immediate family and community-based networks and yet as I noted earlier, we know relatively little about this. We also know relatively little about how digital media such as Skype are being used in our more intimate relationships with partners and lovers in national and transnational contexts.

Rob Kitchin (2014) points out in relation to 'the digital' that there are geographies *of* the digital; geographies *by* the digital and geographies produced *through* the digital. For example, the spatial distribution of Skype is one consideration. A second is the ways in which Skype has altered the nature of social practices and material culture. Third, geographies are now being produced through Skype. Skype needs to be examined not just as part of an empirical research agenda but also in relation to theorisations and methodologies – how we as geographers conduct our own research practice. This study not only involved talking to participants about Skype but also using the technology to conduct a number of the interviews. Before, however, discussing the theoretical concepts underpinning the research (Chapter 2) and the interviews and methodological process adopted (Chapter 3) in the remainder of this first chapter I outline some of the milestones for Skype in order to provide some context for what follows.

Milestones for Skype

Over the past decade the internet has become integral to everyday life for a growing number of people, with more than a third of the global population in 2014 (almost 3 billion) having access to the internet and with over 32 per cent having mobile broadband. Despite this it is important to remember that there is still a digital divide, with 4.3 billion people still not being online, and 90 per cent of these people living in the developing world (International Telecommunications Union 2014, also see Ragnedda and Muschert 2013 on unequal access in internet communication technologies). Gill Valentine (2006, 370) states:

> The Internet expands the opportunities for daily meaningful contact between family members locked in different time-space routines at work, school, travelling, and so on. In this sense online exchanges and daily Internet use are adding a new dimension, rearticulating practices of everyday life and lived spaces.

In relation to Aotearoa New Zealand, where this research was conducted, the trend of people connecting online is similar with internet usage continuing to rise to its current level of 92 per cent of the population (Gibson et al. 2013, 2). This includes emailing, text messaging, tweeting, posting on Instagram and Facebook but also using audio-visual communications services, primarily Skype. In Aotearoa New Zealand 64 per cent of internet users make or receive phone calls online through applications like Skype (Gibson et al. 2013, 11).

Skype was first released in 2003 having been created by the Swede, Niklas Zennström and the Dane, Janus Friis in cooperation with Estonians, Ahti Heinla, Priit Kasesalu and Jaan Tallinn. In September 2005, eBay acquired Skype for $2.6 billion. In 2009, Silver Lake, Andreessen Horowitz, and the Canada Pension Plan Investment Board acquired 65 per cent of Skype for $1.9 billion from eBay, valuing the business at $2.75 billion. Skype, in 2011, was acquired by Microsoft for $8.5 billion. Microsoft's Skype division headquarters are in Luxembourg but most of the development team and 44 per cent of the overall employees of the division are still situated in Estonia (Wikipedia 2015a, Skype.com 2015; also see Skype.com 2014).

Skype software is available to download for free. It enables people to communicate by voice using a microphone, by video using a webcam, and by instant messaging and share file options with other Skype users using the internet. Skype-to-Skype calls to other users are free of charge, while calls to landline telephones and mobile phones are charged via a debit-based user account system called Skype Credit. Skype, in 2015, launched a commercial platform – 'Skype for Business' (formerly Lync) adding to its range and likely revenue. Skype is available on Microsoft Windows, Mac, or Linux, as well as Android, Blackberry, iSO, and Windows Phone smartphones[7] and tablets. It began as one of a number of VOIP (voice over internet protocol) services but is now a mainstream means of communication available for virtually every computer and every mobile operating system. By the end of 2012 Skype accounted for around 25 per cent of all international calls of any kind (Miller and Sinanan 2014). Currently, it dominates the market in the use of webcam for personal communication.

Rapid changes in technology such as enhanced bandwidth and smartphones that connect to the internet are increasingly facilitating services such as Skype and yet the social and cultural changes which have come about through these new technologies are still not necessarily well understood by social scientists. Clearly people have begun to use Skype for a vast array of interactions including family members staying in touch but also for business meetings, job interviews and sexual encounters and much more. A quick online search turns up posts on 'how to keep an eye on your dog when using Skype', 'identical twins stay connected through Skype baking', 'doctor using Skype to hospital in rural Africa', 'Skype plays host to a diplomatic dinner' and 'track Santa with Skype'. There are approximately 40 million users online at peak times (Miller and Sinanan 2014).

In short, Skype offers a variety of services, some for free such as video or voice calling anyone else on Skype, instant messaging and file sharing, and others for a fee such as calls to mobiles and landlines worldwide, sending text messages and Skype for Business. For some specific information around Skype usage, such as average time spent on a Skype conversation and number of people who have ever used Skype, see Table 1.1. Skype often market

Table 1.1 Skype company statistics 2015

Skype Statistics	
Total number of Skype users	74,000,000
Average time spent on a Skype conversation	27 minutes
Number of time that active Skype users spend on Skype per month	100 minutes
Total percentage of small businesses that use Skype as primary communication service	35%
Number of Skype enabled television sets	50 million
Number of iPhone Skype downloads per year	11,500,000
Number of people who have ever used Skype	560 million
Total percentage of Skype calls that are video to video	40%
Average spent yearly by a paying Skype user	$96
Skype revenue in 2010	$406.2 million
Number of monthly log-ins to Skype	124 million
Number of monthly paying Skype users	8.1 million
Amount of money spent by Microsoft to acquire Skype	$8.5 billion

Source: Statistic Brain Research Institute 2015

themselves as 'a community'. Matthew De Beer (2014) wrote on the Skype blog: 'The Skype Community recently passed a major milestone by welcoming its 2,000,000th registered user! We are thrilled that two million people from around the world have joined us to ask and answer questions, share stories and talk about Skype.'

While this research does not focus on Skype as commercial product, more its social and cultural dimensions, it is important to understand some of the financial milestones in its history in order to make sense of how people see this technology as part of their lives. People have a relationship with Skype as a technology, platform, object, filter, orientating device but also via Skype. Sometimes Skype fades into the background, other times (for example, when prompting frustration) it maybe foremost in people's consciousness.

Where to from here?

This introductory chapter 'Why Skype, why now?' began by speaking to the importance of how Skype *feels* – the emotions and affects that this particular communication technology can prompt in people. To be clear, this is not to suggest a determinist relationship between technology or objects and feelings but to acknowledge that online spaces 'matter'. They function to orientate us in particular ways – body to body and body to object. As I have suggested elsewhere (Longhurst 2015), while different technologies, both individually and combined, have the capacity to invoke different emotions I am not implying that these are inherent capacities or qualities that necessarily determine different emotional outcomes (Bingham, 1996, 2005). They do, however, have some role to play (Baym 2010; Benski and Fisher 2013; Clough 2000). Whether media are voice-based, text-based, multi-media, synchronous, or asynchronous can make a difference to the interactions just as real space makes a difference (but again does not necessarily determine) interactions. Therefore, people may well choose carefully which digital media (including various combinations) to use (and sometimes *not* use) in which spaces. This introductory chapter has also provided some milestones or contextual information on Skype, explaining its history and current usage patterns, including in the context of Aotearoa New Zealand where I live and where the research was conducted.

Chapter 2, 'Queer phenomenology: from writing tables to digital screens' explores Ahmed's work on orientation. Given that four of the chapters in the book draw on empirical data to reflect on how Skype might be orientating bodies in particular ways it makes sense in the first instance to orientate myself toward the concept of orientation. The subtitle of this book is 'Bodies, Screens, Space'. Ahmed (2006, 22) remarks that she turned her attention in *Queer Phenomenology* to the philosopher Husserl's table (as in a piece of furniture rather than a set of information arranged in rows and columns): 'Once I caught sight of the table in Husserl's writing, which is revealed just for a moment, I could not help but follow tables around.' I feel like I have been following around bodies, screens and space for the past couple of years. Each of these – bodies, spaces, screens – is discussed in Chapter 2 interwoven with ideas inspired by Ahmed's writing on queer phenomenology and a *Cultural Politics of Emotions* (2004) in which she discusses bodies taking the shape (or not) of the contact they have with objects and others. While I had already read some of Ahmed's work prior to conducting the fieldwork, I did not set out thinking that I was going to necessarily draw on it as heavily as I have. Ahmed's queer phenomenology began to spark thoughts for me only when I was analysing material and writing draft chapters. As people started to tell me their stories about using Skype I mulled them over and was struck by how comfortable some were in the spaces they were inhabiting for particular purposes and how uncomfortable others were.

After examining queer phenomenology, I explain how the research was conducted in Chapter 3. Data were derived from interviews with 39 participants (25 women and 14 men). While some participants were interviewed face to face others were interviewed via Skype. Digital media in recent years have opened up new methodological possibilities and literature is beginning to emerge on this topic. Much of it though pays attention to the many pragmatic issues involved in using online methods, for example, design, building rapport and ethical considerations. In Chapter 3 I report on what participants told me about how they *felt* about being interviewed face-to-face or via Skype. Using the medium to conduct a number of interviews meant I was able to fold the methodological process and data into each other. This chapter both reports and reflects on how interview data, and also information from online sources (such as searches using Google and Reddit) were collected.

Chapter 4, 'Selves, others, objects and space' addresses a number of what are everyday, but vitally important, issues surrounding Skype including seeing ourselves in the box in the corner of the screen, what we feel and think when we see others, including their background. Sometimes people consider carefully what might be seen in their background, other times not. Sometimes people show us objects or rooms. Sometimes they may be out in the public sphere talking to us on a mobile device when we Skype them. This chapter canvasses research participants' thoughts on these matters before moving in the next chapters to address more specific issues.

Chapters 5, 6 and 7 focus on the three broad themes of family, work and sexual intimacy. In Chapter 5 it is argued that family, friends and loved ones are using Skype to establish and maintain relationships across physical distances. This is especially the case for grandparents and grandchildren. A number of family members, friends and loved ones have now been using Skype for several years, meaning they are now less self-conscious about the technology and their primary concern is tending to the relationships and emotions of those involved. The chapter focuses on intergenerational connections, using Skype for special occasions, developing a sense of space using Skype, and how most participants reported feeling comfortable in these interactions. Online and offline spaces coexist reasonably seamlessly, enabling family, friends and loved ones to stay connected. Bodies and spaces – cyber and real – are happily entangled. Or, drawing on Ahmed (2004 and 2006) work on objects, orientations, space, time, directionality, bodies and proximities we could describe the bodies discussed in this chapter as feeling orientated and able 'to sink'[8] into the spaces they are inhabit through Skype.

Chapter 6 'Skype for work: "a bit weird"' illustrates that people are less comfortable 'sinking' into work spaces using Skype, especially for job interviews. At work there appears to be a feeling that there is more at stake if the technology fails, if the body is not well-presented, if the background during the Skype call reveals more or less than is anticipated by either party. People are not yet as used to or as familiar with using Skype for the purposes of paid employment as they are for the purposes of connecting with family, friends and loved ones. In many cases it disrupts in an uncomfortable way the binary division between work and home and between private and public space. It seems that in time we may grow used to this and begin to feel more comfortable using Skype at work but for now the bodies and objects, to use the phrase of Ahmed (2006), do not easily 'line up', they are 'oblique'.

In Chapter 7, 'Skype Sex: "Queer Effects?"' it is argued that nearly all of the participants understand sex via Skype as contrived or unnatural, a queer effect of the audio-visual that risks deviation from the 'straight and narrow'. Skype appears to mediate an act that they consider ought to be about a connection between real bodies, in offline space. Skype, they argue, produces an artificial version of the real, or a queer version of the straight. Most think it is a poor substitute for real co-present intimacy.

Chapter 8, 'Reorientating bodies and spaces' returns to Ahmed's queer phenomenology to consider how moments of disorientation might be important (but not always) for prompting people to question their own beliefs, the ground on which they are situated and reside. Does Skype at this historical moment, while it still feels strange in some contexts, offer an opportunity to rethink subjectivities, bodies, relationships with objects, others and space. What can ordinary moments teach us? And, who is 'us'? Ahmed (2006, 159) notes that 'disorientation is unevenly distributed: some bodies more than others have their involvement in the world called into crisis'. Does Skype work to privilege some bodies at the expense of others? These questions are addressed in the concluding chapter. The aim is to bring together the various lines of enquiry and to make sense of them. In the first instance though it is important to outline the theoretical constructs which underpin arguments made throughout the book. This is the work of the chapter that follows.

Notes

1 Aotearoa is the Māori term for what is commonly known as New Zealand. Since 1987 when the Māori Language Act was passed making Māori an official language, the term Aotearoa has been used increasingly by various individuals and groups. For example, all government ministries and departments now have Māori names which are used, in conjunction with their English names, on all documents. In some instances both Aotearoa and New Zealand are used together. In other instances one

or the other are used separately. Throughout this book I mainly use the term Aotearoa New Zealand to acknowledge the bicultural nature of the nation but it is important to highlight the point that the naming of place is contestatory (Berg and Kearns 1996; also see Larner and Spoonley 1995).

2 Throughout this book I use the term 'bodies' to refer to human bodies. Geographers have, however, questioned the binary division often drawn between humans/animals, culture/nature (Anderson 1995) and more recently, inorganic/organic (Kinsley 2013a). Another point worth making in relation to human bodies is that in the early 1990s Moira Gatens (1991) argued that there is not *one* body – *the* body is a masculinist illusion. There are only bodies in the plural. Much feminist discussion focuses on the complex processes through which female and male bodies are differentiated, sexed and gendered. They are also racialized, marked by class, are a particular shape and size and so on. This point remains as relevant today as when it was made more than 20 years ago.

3 Many geographies of feeling and 'emotional life' (Thien 2011) are rooted in feminist geography (Davidson and Bondi 2004, Davidson and Milligan 2004, Davidson et al. 2005, MacKian 2004; Parr 2005; Thien 2005; Tolia-Kelly 2006). There have been numerous debates about the relationship *between* emotion and affect with some (often feminists) placing more emphasis on feeling and emotion linking it to the individual and personal while others have preferred to employ the term affect linking it to the social, transhuman and political. Still others (Bondi and Davidson 2011; Sharp 2009; Wright 2010) prefer to focus on the interconnections between emotions and affects, examining how they are relational and move through bodies and spaces. This is the perspective adopted in this research.

4 Media theorists used to discriminate between technologies or media (e.g. mobile phones and computers) and platforms (e.g. email and landline voice calls) but now that some platforms can be accessed via different technologies there has been a merging which 'makes it difficult to retain categorical distinctions ... given that all these continue to hybridise and overlap' (Madianou and Miller 2012, 104).

5 On mothering and technology also see Gatrell (2008) on women's use of the internet; Lim and Soon (2010) on mothers in China and South Korea utilizing digital media to fulfil their maternal duties of managing homes and families; Cassidy (2001) on 'Cyberspace meets domestic space'; Lim (2008) on introducing technology into a household; Flanagan (2000) on gendered concepts being embedded in the construction of online worlds; Silverstone, Hirsch and Morley (1992) on digital media and the moral economy of households; and Koerber (2001) on 'feminist mothers on the web'.

6 I have used scare quotes around the word real here to signal the point being made, that is, that it is not possible to separate out the real or reality from representation. Signs do not just name the world but construct it in particular ways. Reality and representation are mutually constructed rather than separate. I do not continue to use scare quotes around the words real and 'reality' throughout the text but I do continue to bear this point about the entanglement of the real and representation in mind.

7 Smartphones typically function as phones but also as portable media players, cameras, videos, GPS navigational systems and a variety of other functions. They have high-resolution touch screens and web browsers that can access and effectively display standard web pages. 'Skype has been downloaded approximately 200 million times on iPhone and Android phones' (Miller and Sinanan 2014, 3).

8 As noted in Longhurst (2013) I owe thanks to Tracey Skelton for alerting me to the idea that bodies not only need to 'sink' (Ahmed 2014) into the spaces of Skype but they also often need to 'sync' in time with calls needing to be 'lined up' across international time zones through Skype.

2 Queer phenomenology

From writing tables to digital screens

Phenomenology, with its emphasis on lived experiences, intentionality of consciousness, positioning in relation to objects and habitual actions in shaping bodies and worlds, has been practiced in various guises for centuries although arguably it has come into its own in the early 20th century in the works of Husserl, Heidegger, Sartre, Merleau-Ponty and others. Human geography, like many other disciplines, has also been influenced by phenomenology. In the 1970s and 1980s humanist geographers reflected on people's lifeworlds (Buttimer 1979; Seamon 1979; Tuan 1974; also Rodaway 1994 for a later example). In the 1990s, continuing into the 2000s, non-representational geographers began to consider embodiment, performativity and spatiality (Anderson and Harrison 2010; Bell et al. 1994; Lorimer 2008; Thrift 2008). More recently there has been a turn towards what some are calling a distinct post-phenomenological way of thinking (see Ash and Simpson 2014 on 'geography and post-phenomenology'). There is no shortage of rich work, both historical and contemporary, in human geography that has drawn on insights from phenomenology. My own work on the body has undoubtedly long been influenced by this scholarship. It is, however, phenomenological work inspired by feminism (e.g. Marion Young 1990) that really ignited my passion from early on in my career.

Getting orientated

For decades feminist scholars, including feminist geographers, have argued convincingly about the importance of understanding themes such as homes, dwelling places and politics of location in relation to gender (Davidson 2001; Grosz 1990, 1994; Haraway 1991; Rich 1986; Young 1990). In particular, in recent years I have been indebted to Ahmed for her work on 'queering phenomenology and moving queer theory towards phenomenology' (Ahmed 2006, 5). Ahmed (2006, 161) explains: 'To make things queer is certainly to disturb the order of things ... the effects of such a disturbance are uneven, precisely given that the world is already organised around certain forms of living – certain times, spaces and directions.'

Ahmed's musings on space appeal enormously to my sensibilities as a geographer, particularly one who has long been interested in studying sex, gender and the body. I am especially compelled by her research on queer phenomenology (Ahmed 2006) in which she considers carefully the concept of orientation in and of phenomenology. Thinking through the object of the philosopher's table Ahmed skilfully explores themes of space, time, directionality, bodies and proximities. About a year ago I became intrigued by what her queer phenomenology might be able to bring to the table for a social, cultural and gendered geographical study of Skype.

Ahmed is Professor of Race and Cultural Studies at Goldsmiths College, University of London.[1] Her publications are far too numerous to mention here. A list I compiled of books, edited books, book chapters, journal articles, special issues of journals, interviews and columns totals over 100 but her authored books include *Queer Phenomenology: Orientations, Objects, Others* (2006), *Differences that Matter: Feminist Theory and Postmodernism* (1998), *Strange Encounters: Embodied Others in Post-Coloniality* (2000), *The Cultural Politics of Emotion* (2004), *The Promise of Happiness* (2010a), *Feminist Killjoys (And Other Willful Subjects)* (2010b), *On Being Included: Racism and Diversity in Institutional Life* (2012), *Willful Subjects* (2014) and *Living a Feminist Life* (forthcoming). Themes addressed in these books and Ahmed's other work include feminism, Otherness, emotion, home, racism, diversity, whiteness, happiness, queer politics, skin, postcolonialism, affective economies, love and many more.

I suspect that it was never Ahmed's intention that her work be used to consider Skype and yet many of the concepts she explores especially in *Queer Phenomenology* are useful for thinking through people's bodies and their relationship with screens, images and voice. In her research on 'mixed orientations' (mixing races) she notes it is 'not simply about what a body has, or even what a body can do, but involves a material and affective geography: affecting the way we gather: bodies, objects, worlds that come together as well as break apart' (Ahmed 2014, 94). While Ahmed in this particular contribution addresses racism and how whiteness is produced, her ideas have helped me to consider how people currently inhabit different spaces through Skype, how different orientations of the body, screens, images and voice might affect relationships with family, friends, lovers and colleagues. This is not to suggest that Ahmed ignores technology. She explains that:

> Technology does not simply refer to objects that we use to extend capacities for action. Technology (or *techne*) becomes instead the process of "bringing forth", or, as Heidegger states, 'to make something appear, within what is present, as this or that, in this way or that way' (159).
>
> (Ahmed 2006, 46)

Ahmed has a long history in her philosophical work of addressing social and political questions. She is interested in gender, sexuality and race and combines the philosophies of the some of the most well-known figures such as Husserl, Heidegger, Merleau-Ponty and Fanon with critical race theory, feminist theory and queer studies. This integral aspect of her work I have also found to be deeply important since research on digital technologies, media, and representation needs to be situated within a social, cultural and political framework.

I am certainly not alone in thinking this. In an online call for papers in 2015 for a special issue on 'Digital Technologies and Social Transformations: What Role for Critical Theory?' guest editors Delia Dumitrica and Sally Wyatt explain that over the past two decades social research on the role of digital technologies in contemporary transformations has emerged as a specific disciplinary field in its own right (Dumitrica and Wyatt 2015). In the process there has been a move away from overly positive and enthusiastic accounts of the potential of these technologies to address social problems and alleviate inequalities, towards a more critical and careful understanding of the mutual constitution of digital technologies and social structures. Calls for recovering the role of critical theory in furthering thinking about digital technologies (Feenberg 1991, 1999; Feenberg 2009a, 2009b, Longhurst 2009) have high-lighted the importance of developing suitable theoretical tools. While this book presents an array of empirical data it is also about pushing in new directions the work of theorists who focus on queer, bodies, screens and space to deepen understandings of people's use of Skype at this particular historical juncture.

Although I have focused attention on Ahmed (2006) *Queer Phenomenology*, literature from a vast array of disciplines and sources has informed the project. At times I imagined myself to be like a magpie collecting gems from disciplines that are not my own, taking them to use for improper and unintended purposes. I certainly did not stay within the disciplinary confines of human geography in conducting this project, nor did I stick to academic literature. I read widely, pulling out what I thought were interesting contributions from others that I could use to make sense of Skyping. I went beyond articles and book chapters searching the internet (blogs, forums, advertisements) on topics such as 'Skype sex', 'job interviews via Skype', 'Skype weddings' and 'using Skype in the classroom', bringing these materials together in the pages of this book.

The reading I have done falls broadly into the three categories – bodies, screens and space. Woven throughout categories or themes, however, are insights from queer theory particularly, as previously mentioned, Ahmed's Queer Phenomenology (2006) but also Eve Kosofsky Sedgwick's Tendances (1993), Judith Butler's (1990) *Gender Trouble* and a variety of others. Ahmed (2006, 22) explains that she feels it is risky reading philosophy as a non-philosopher: 'When we don't have the resources to read certain texts, we risk getting things wrong by not returning them to the fullness of the intellectual histories from which they emerge. And yet, we read.' I have attempted to 'take care' in the reading I have done (as Ahmed suggests) not just of her work and the work of other philosophers but also of blogs, websites, and tweets. All these works matter, albeit in different ways, in the production of knowledge about Skype. I acknowledge though that others in disciplines beyond my own may not necessarily agree with my 'take' on things. As Ahmed (2006, 22) notes: 'Disciplines also have lines ... Such lines mark out the edges of disciplinary homes, which also mark out those who are "out of line"'. It is likely that there will be those both outside but also inside my own discipline of human geography who feel this project of using queer phenomenology to inform a project on Skype is 'out of line'.

Spinning outwards

The term queer, as a critical theory and methodology, originates in gay and lesbian studies. Queer has provided a platform from which to critique hetero-normativity (Sullivan 2003). Yet, as Eve Kosofsky Sedgwick argues queer theory can be usefully applied to a broader range of normative knowledges and identities than just sexual ones. She explains:

> a lot of the most exciting work around 'queer' spins the term outward along dimensions that can't be subsumed under gender and sexuality at all: the ways that race, ethnicity, postcolonial nationality criss-cross with these and other identity-constituting, identity-fracturing discourses, for example.
>
> (Sedgwick 1993, 8–9, italics in original)

By way of an example, one project that spins the term queer outwards is Fiona Giles' (2004, 301) examination of 'breastfeeding behaviours [that sit] outside the normative constraints that apply in contemporary Western culture'. She refers to these behaviours as 'queer breasting'. Giles (2004, 301) explains the 'contemporary norm is that between 50–90 per cent of mothers in industrialized nations will begin breastfeeding their babies before leaving hospital, and wean them within the first year'. Some breastfeeding, however, sits outside this norm. The aim of Giles's book is to examine the social meanings of breastfeeding (as opposed to the nutritional content or medical benefits of 'mother's milk') which have not to date been paid much academic attention. She covers subjects such as cooking with breast milk, donating milk to a milkbank after the death of a child, adult nursing, and lactation pornography as well as some of the more rehearsed subjects such as weaning an older child and mastitis. Well known popularist writer on pregnancy and birth, Kitzinger, claims Giles' book says: 'All the things that the other books about breastfeeding don't say' (front cover).

Giles's (2004) work on 'queer breastfeeding' prompted me to draw on queer theory to inform research on women's experiences of work (Longhurst 2008). Again, this is not a topic that one might expect to be underpinned by queer theory. To quote Sedgwick (1993, 8) again, it: 'spins the term outward along dimensions that can't be subsumed under gender and sexuality'. Concepts such as queer and others related to it such as 'the closet' (which I used to discuss women hiding being pregnant at work) helped me think through some of the spaces – material, discursive, and imaginary – and power relations surrounding the bodies inhabiting these spaces.

By way of one final example, I also used queer theory as a lens thorough which to view fat or large bodies (Longhurst 2014).[2] By un-fixing or queering fat bodies the intention was to illustrate how body size and shape are disciplined by social institutions and practices that normalize and naturalize just one type of body and shape over and above others. Jagose (1996, 16) argues: 'heterosexuality is too often represented as unremarkable'. In other words, it is treated as the normative position from which other sexual identities are thought to deviate. Following this line of thought it can be argued that slimmed bodies are also too often represented as the unremarked norm from which other body shapes and sizes are seen as derivative – fatter or thinner than the norm, taller or shorter than the norm, and so on. In fact, many researchers have built on this idea of 'spinning the term outward' and have queered dimensions of subjectivity which might on first appearance seem surprising to some.

Natalie Oswin (2008, 90) makes the point that new scholarship in queer studies 'merges postcolonial and critical race theory with queer theory to bring questions of race, colonialism, geopolitics, migration, globalization and nationalism to the fore in an area of study previously trained too narrowly on sexuality and gender'. Queer theory can also merge other theories to bring questions about pregnancy, body size, digital media and a variety of other topics to the fore. For example, geographers have combined spatial theories and queer theories to focus on tourism (Johnston 2005, also see Johnston (2015) for a review of 'genderqueer geographies'). To reiterate Sedgwick's (1993) point, although queer studies has mainly been used as a platform from which to critique heteronormativity, it can also be applied to a broader range of normative knowledges and identities than just sexual ones.

Perhaps then, given that it is possible to queer many things, not just knowledges and identities but also processes, interactions and spaces, the question ought to be, why not Skype? Jackie Wykes (2014, 4, italics in original) in relation to queering 'fat' bodies argues:

> queer can be either (or both) a description *and* an action; an orientation *and* a practice; a mode of political and critical enquiry which seeks to expose taken-for granted [sic] assumptions, trouble neat categories, and unfix the supposedly fixed alignment of bodies, gender, desire and identities. The term queer's definitional heart or originary meaning is 'same-sex sexual object choice'.
>
> (Sedgwick 1993, 8)

While I am not suggesting that the term be displaced so far from this originary meaning that it may strip away the possibilities and power of queerness itself I do think it can be used to broader affect than focusing just on gender and sexuality.

Queer provides fertile ground for a range of interdisciplinary engagements on a broad array of topics. The conversations that we have do not represent one-way relationships and ideas and once transferred from one disciplinary area to another would not remain unchanged. For example, human geography or digital humanities may offer up just as many important insights to those in queer studies as queer studies may be able to offer up to those in human geography or digital humanities.

It is important to remember, however, that queer studies is not a uniform body of work and some threads may be more productive than others for examining people's relationship with Skype. I have chosen to examine primarily Ahmed's concept of orientation and disorientation through the lens of phenomenology. Disorientation, explains Ahmed (2006, 160) 'occurs when we fail to sink into the ground', when the ground itself and objects are disorientated. Ahmed uses the term queer in two senses, first as a way of describing what is 'oblique' or 'off line' (and may therefore have the effect of queering space), and second to describe specific sexual practices. For the most part in this book, I adopt Ahmed's first meaning of the term queer. This enables me to address the question: is Skype queering space? In Chapter 7, however, the focus rests on Skype sex, and because some of these practices could arguably be described as non-normative, I adopt Ahmed's second meaning of the term queer. Others may well find value in using different meanings and concepts from queer theory to understand embodied communication in digital environments. It is a rich and diverse field.

Bodies

Bodies are an important site for understanding how power operates since it operates differently on and through different bodies (Braziel and LeBesco 2001; Pile 1996; Probyn 2000; Rose 1993; Teather 1999). As feminist and other critical scholars have been arguing for decades, bodies are not simply human bodies but are sexed, gendered and racialized (Davis 1997; Gatens 1988, 1991). They are also a particular age, size, shape, and class (Duncan 1996; Johnson 1989; Johnston 1997; McDowell and Court 1994; Moss and Dyck 2002; Nelson 1999; Pratt, G. in collaboration with the Philippine Women Centre 1998). Bodies are also intertwined with space (Probyn 2003). The materiality of bodies (whether they be pregnant, fat, brown, white, disabled, straight) is always located (Butler and Bowlby 1997; Butler and Parr 1999; Longhurst 2001; Moss and Dyck 2002; Nast and Pile 1998) and interpellated through a range of discourses. Butler (1990) in discussing this interpellation emphasizes the role of repetition in bodily performativity, making use of Jacques Derrida's (1988) theory of iterability – a regularized and constrained repetition of norms This repetition, she argues, is not performed by a subject; rather it is what enables a subject and constitutes the spatial and temporal condition for the subject. Rachel Colls (2007) in 'Materialising bodily matter' draws on Butler (1993), Karen Barad (2003) and a range of other theorists to extend the notion of performativity by focusing on matter, the material and materialization (her particular focus in on fat bodies). Colls (2007) argues that performativity needs to grant not only language and culture agency and historicity but also matter itself needs to be understood as having productive capacities. Performative bodies are discursive but constantly lived through their materiality in particular times and spaces.

What happens when this materiality is, for many interactions, filtered through a screen that enables seeing and hearing but not touching (Brown et al. 2011), tasting and smelling (Stone 1991)? Brian Massumi (2002) argues that cultural theorists need to pay more attention to the most significant characteristics of embodied existence, that is, movement, affect and sensation in relation to media. Do movements and senses when 'assembled' (Anderson et al. 2012) through different media change our 'emotional geographies'? (Anderson and Smith 2001; Pile 2010a). Is Skype, in filtering bodies through screens, helping prompt 'trouble', including 'gender trouble' (Butler 1990) in a different kind of way? Are we tending to perform our bodies on Skype repeatedly so that gestures and positions (most often camera shots of heads and shoulders) have become the norm, seemingly the natural way to 'do' Skype and if so what might be some of the consequences of this view?

Over the past few years I have come to think much more about how bodies are interpellated through various digital media. People live different and intersecting (sometimes contradictory) subjectivities in different and intersecting spaces including online spaces. Bodies both produce and are produced by these both offline and online spaces. Bodies and spaces collide, rub, dissolve (see Davidson 2001 and 2003 on agoraphobia) and fold into each other. When screens are involved it seems that maybe bodies may collide, rub, dissolve and fold differently.

Continuing to think about bodies, it is useful to turn attention to the concept of orientation. Ahmed (2014, 95), drawing on the work of the philosopher Edmund Husserl, argues: 'Orientations are about how we begin, how we proceed from "here"'. The differences between here and there matter. They orientate our bodies in particular directions. This, it could be argued, is especially the case in relation to Skype. It matters that I am here and you are there – physically removed from me but visible through a screen. I can see and hear you (at least all going to plan) but I cannot touch, taste or smell you. Our worlds unfold in a sensorially different way than in the past. People and objects appear in front of me but it is now the image on the screen that I am increasingly becoming used to. The computer enables not only sound but also 'real-time' visual images to be transmitted across space and time. People have long been able to keep 'in touch' via phone which transmits voice but Skype with video allows for visual images to be transmitted. These images may have the potential to prompt different feelings of proximity (distance and closeness) between people (e.g. see Valentine and Holloway 2001a on rural children's use of the internet, which offers them a 'window on a wider world').

Typically on Skype people are visible from the waist up only. I see their face in front of me. Sometimes the frame freezes and I see a particular (often unflattering) image frozen on the screen. Sometimes I see things – objects, images, interactions – in the background which are unexpected. The person on the screen is not always aware of them. In the future it is likely that people will become increasingly comfortable with Skype as a communication technology, leaving it on for longer periods and getting up and moving around during the call. It is likely that in time it will be embedded as an everyday and accepted part of people's lives (Miller and Sinanan 2014) and this means that we will start to perform our bodies differently as part of the interaction.

Scholarly contributions on Skype to date, however, have not tended to consider in-depth this bodily performance as part of the communication. The framing of bodies as an interaction is important though because it is imbued with power relations. It does not determine the outcome of interactions but neither is it benign. Space, both offline and online, matters. Ahmed (2006, 1) poses the question: 'What difference does it make "what" we are orientated toward?' She adopts a phenomenological model of bodies and emotions as being 'directed' towards objects, arguing: 'Emotions are relational' (Ahmed 2004, 8). There are a number of different approaches to emotions. Some theorists see them as tied primarily to bodily sensations (a view often ascribed to Descartes) while others see them as being tied to cognition and judgment (a view often ascribed to Aristotle). Still others see them as being about sensation and cognition. Ahmed builds on these various theories by arguing that emotions are shaped by contact with objects. She explains: 'Emotions are both about objects, which they hence shape, and are also shaped by' (Ahmed 2004, 7).

While Ahmed is focused primarily on sexual and racialized orientations her argument is also useful for considering more generally how people 'reside in space' (Ahmed 2006, 1). She notes that by foregrounding orientation it is possible to retheorize the sexualisation and racialisation of space and spatiality of sexual desire and race politics. It is also possible to retheorize the use of what is becoming a routinely used technology in many people's lives, for example, Skype. Putting human geography, queer studies, phenomenology and media studies in closer dialogue offers a new way of thinking about the spatiality of our everyday embodied interactions. Ahmed (2006, 2) uses phenomenology because it makes 'orientation' central to the argument that 'consciousness is always directed "toward" an object'. This offers a useful way of thinking about the *spatiality* (both online and offline) and visuality of contact as constituted via Skype.

Ahmed (2004a, 1) explains: 'Bodies take the shape of the very contact they have with objects and others.' She says: 'words are not simply cut off from bodies, or other signs of life. The work of emotion, then, involves the "sticking" of signs to bodies' (Ahmed 2004a, 13). Ahmed (2004a, 152 (see also Ahmed 2004b, 2004c)) notes:

> Comfort is the effect of bodies being able to 'sink' into spaces that have already taken their shape. Discomfort is not simply a choice or decision – 'I feel uncomfortable about this or that' – but an effect of bodies inhabiting spaces that do not take or 'extend' their shape.

Given my interest in the body, especially its fleshy materiality and spatiality, it seems that phenomenology, with its emphasis on lived experience, can offer useful insights. Ahmed (2006, 2) explains that she 'arrived at phenomenology because, in part, the concept of orientation led' her there. She explains that bodies take shape as they move through spaces directing themselves toward or away from objects and others. This can help us think about relationships with computers, mobile devices, webcams, speakers, cables and screens, these being the main objects under consideration in this book. Ahmed argues: 'consciousness is always directed "toward" an object' (2006, 2). Through objects people are able to reconfigure relationships that create different emotions and affects (Brennan 2004). Being 'orientated', Ahmed explains, means feeling at home, knowing where we stand, or having certain objects within our reach. For example, many people today feel uncomfortable when they do not have their phone, computer, iPad or some other device at hand or within reach. These, for some, have become objects that need to be proximate. So, while Ahmed pays attention to queer orientations, it seems these ideas are also applicable to bodies in relation to people's use of different digital media including Skype.

Ahmed argues that queerness disrupts and reorders relations of proximity by not following accepted paths. For some using Skype might cause disruption, for others *not* being able to use Skype might cause disruption. There is also the question of exactly what Skype is being used for. For example, it may reorder relations of proximity in regard to sexual desire but not catching up with family members. Some things appear more comfortable via a screen than others.

Screens

Having a screen in between ourselves and others does not immediately imply an added mediation although some have argued this case (Turkle 2011). I tend to be more persuaded by the idea that: 'there are no unmediated, pure relationships. All the ways in which relationships exist, including communication, are cultural activities' (Miller and Sinanan 2014, 3–4). There is little doubt, however, that new media are likely to be first experienced as an extra mediation. The screen that transmits images may be perceived as robbing the relationship of something real, immediate and true that two or more people standing in the same room may share. Miller and Sinanan (2014, 6) argue that there is often a lament for a past authenticity that predates the emergence of new digital technologies that resonate 'with popular assumptions which are almost universally held, and constantly reinforced in journalism. New technologies are making humanity itself more artificial and thereby less intrinsically human.' These popular and academic discourses, argue Miller and Sinanan, are highly conservative. What is argued in this book is not that Skype is adding a further mediation to communication but that what we are experiencing is our bodies being orientated differently through screens providing different frames on the world, some of which may make us feel uncomfortable, others more comfortable. Comfort levels around these new orientations are inseparable from the bodies under consideration.

Existing work on digital media covers the use of Skype by a range of different individuals and groups, including people who are deaf and hard of hearing (Valentine and Skelton 2008), those on the autistic spectrum or suffering from anxiety disorder (Davidson and Parr 2010), children and young people (Holloway and Valentine 2001a, 2001b, 2001c; Valentine and Holloway 2002), those seeking romance and sex (Baker 2000, Ben-Ze'ev 2004, Johnston and Longhurst 2010; McLelland 2002, Schaeffer-Grabiel 2004 on cyberbrides), community groups (Bryson 2004 on 'queer women on the net'), and those who like to engage in online games (Shapiro 2010 on *Second Life*, Woleslagle 2007 on *EverQuest* and Valkyrie 2011 on Massively Multiplayer Online Role Playing Games more generally). It is demonstrated in this considerable body of work that the internet and screens are not spaces that can be separated or disconnected from the offline spaces (Kitchin 1998a, 1998b; also see Adams 1997, 2007; Crang et al. 1999). For example, while the internet enables child-mother-grandmother relations to stretch over longer distances and different time zones, and to continue intimate and familial relationships long after children have left the parental home it does necessarily change existing gender roles and relations (Vancea and Olivera 2013). Mothers continue to typically run households and take responsibility for care-giving. As Madge and O'Connor (2005, 83) argue there is a 'simultaneity of online/offline worlds' in the lives of women who use the internet as part of their mothering practices. They explain that online and offline spaces coexist and that 'although cyberspace[3] can result in the production of new selves, these selves have residual attachments to embodied experiences and practices' (Madge and O'Connor 2005, 83).

Despite this simultaneity of online and offline worlds many still believe that screens add a layer of mediation that somehow creates a boundary separating the real and the virtual. It is perhaps therefore worth pausing to think a little more about this before reflecting in more detail on space. Screens provide us with visual images (which may be blurred, stilted, pixelated, frozen, or crystal clear) of those we connect with. People and things are projected on the screen and most of us (but not all) use our eyes to read and interpret these images. In fact, the world that most of us (but again not all) engage in is highly visual. Despite this, as Gillian Rose (2001, 2) argues, 'it still isn't clear exactly what that [encountering the world visually] might mean'. Therefore it is important to continue to canvas ideas about the visual and bodily interactions with computers and other devices. Rose and Tolia-Kelly (2012, 1, italics in original) state that it is important to 'attend to the relationships *between* the "visual" and the "material"', and 'explore what kinds of new thinking might emerge in that intersection'.

Over the past few years as I have been conducting research on digital media including Skype, I have heard a number of stories about people, but especially children, touching the image of their friend, loved one or family member on the screen. I was told in casual conversation one evening about a toddler who, when she sees her grandmother on the screen, gets really excited about talking to her, and about showing her drawings, and sometimes kisses the computer screen (see Figure 2.1).

Figure 2.1 One-year-old grabs for the mobile phone when he sees his father on the screen

Ahmed (2006, 51) poses the question: 'How do bodies "matter" in what objects do?' Objects – whether they be drawings or screens – enable us to do particular things but not others. The screen, orientated in a particular way, may enable us to see an image of our loved one but not touch or taste their skin or smell the scent of their body. In this way, as Lefebvre (1991, 143) suggests, 'space "decides" what actually may occur'. People have long inhabited space with objects but increasingly one of those objects is a computer, or mobile device with a screen. This co-dwelling of bodies and screens warrants further consideration – how bodies move around and in front of screens is important and likely to change as people become more familiar with different technologies. Ahmed (2006), calling on the work of Husserl (1989) says that bodies and objects become more than simply 'matter' insofar as they can make an impression. Bodies can make an impression *visually* on screen, and aurally through computer or device speakers, but the other senses including touch are absent in this impression – there is no pressing of skin to skin, although there may be skin to device and skin to screen. Husserl (1989, 155) states that bodies are 'something touching which is touched'.

In the case of Skyping this touch is not felt on or through the skin of the body but instead the hard surface of the screen. Screens vary. Some simply display information through a liquid crystal light display while others, known as 'touchscreens', are display devices that also take input via embedded pressure sensors. To screen or screening (as a verb) means to select, filter, separate or partition. Screens effectively filter skin-to-skin touch. Lynda Johnston (2012, 1) in writing about 'the spatial politics of touch' in relation to drag queens notes:

> Sensuous experiences of touching, being touched, and embodied feelings associated with touch can ... provide an understanding about people's sexed and gendered subjectivities. When bodies touch it is the closest they can be, in the same place, and at the same time.

If touch in this way – skin-to-skin – is how people are used to experiencing their closest moments then Skype is likely to reconfigure this, to reorientate the bodies in space. 'Haptic geographies – bodies that touch places, places that touch bodies, and bodies that touch each other' (Johnston 2012, 1) through screens is and will continue to change. How this is the case, at least in part, is the subject of the remaining chapters.

In 1994 when Paul Rodaway published *Sensuous Geographies: Body, Sense and Place*, which focuses on the different body senses – touch, smell, taste, aural, sight – it was unlikely that most of us were envisaging quite how much our everyday worlds might be changed by our senses being filtered through screens. Screens as objects can be understood as assemblages of matter. They have the capacity to shape (but not wholly determine) emotion and affect. James Ash (2015a, 84) suggests: 'To understand the affective capacities of technology, one should understand how technologies reorganise and draw upon associated milieus' [sic] to generate affect and how the material thresholds of objects shape what these affects are.' The organic and inorganic co-mingle.

Paterson (2006, 691) also examines this co-mingling of technologies and bodies conducting research on 'technologies of touch and distance'. He argues: 'Although physical touch is often associated with proximity and intimacy, technologies of touch can reproduce such sensations over a distance, allowing intricate and detailed operations to be conducted through a network such as the Internet' (Paterson 2006, 691). A haptic device that comes to mind in relation to touch and use of Skype is the 'Kissenger' (referred to in Chapter 7 on Skype sex), which has been designed over the past few years as a physical interface (a pair of robots) that enables kissing through digital media. Paterson (2006) argues that the implications of these new technologies are significant, as they reconstruct the human-computer interface from being primarily audio-visual to being much more multisensory.[4]

Paterson and Dodge (2012) also address touch and touching, asking why it has been largely ignored by social scientists for so long given that it is so central to everyday embodied existence. They are interested in the place of touch in a variety of spaces including spaces of care, work, creativity, recreation sociality and domesticity. It is perhaps not surprising therefore that their edited collection covers an array of topics including animal encounters, tourism, massage, beauty treatments, professional medicine, everyday spiritualities, touch free spaces of automated toilets, scholarly fieldwork and paintings of fleshly bodies. There is little, however, in the collection on touch (or lack of) through screens.

Cranny-Francis (2013) on the other hand focuses directly in *Technology and Touch* on exploring the development of various new touch technologies, both technologies that people reach out to touch (such as laptop computers, iPads and mobile phones) and technologies that touch people (such as smart clothing, wearable computing devices and robots) by exploring how we use touch to connect with and understand our world, and ourselves.

In these two sections on bodies and screens I have also discussed space on account of the fact that it is difficult to separate out space from bodies and screens. In the section that follows, however, I raise a couple of additional specific points in relation to online space which I think are worth reflecting upon before moving in the next chapter to discuss methodology.

Space

The topic of space has long been on the agenda in the humanities and social sciences but recently it has regained some prominence. Couldry and McCarthy (2004) explain there has been a 'geographical turn' in media studies and it has been recognised that a deeper consideration of the various forms of spatiality created through mediation is required (also see Wilken and Goggin 2012). This so-called geographical or spatial turn, prompted in part by geographers, has allowed scholars in a range of other disciplines including digital humanities, queer studies, feminist studies, gender studies, history, anthropology, sociology, and philosophy to name just a few to return to the important, but for a period ignored, question, what is space? The productive debates that emerged have led to a proliferation of literal and metaphorical spatial references. Terms such as 'boundaries', 'areas', 'location', 'state', 'terrain', 'site' and 'region' have become part of many scholar's lexicons. Added to these terms are others often associated with online spaces such as 'cyberspace', 'virtual space', 'imagined territories', 'spaces in online games', 'Twitter', 'Facebook' and 'Skype' (see Graham 1998 on 'conceptualising space, place and information technologies'). This proliferation of terms signals an intellectual enrichment of thinking about space (e.g. see Farman 2012 on how mobile technologies are changing the ways people produce lived, embodied spaces). It also means, however, that how space is understood, including the spaces of social media, has become increasingly equivocal, multifaceted and at times incongruous (e.g. see Parr 2002 on the embodied spaces of health and medical information on the internet).

Sam Kinsley (2013a) examines digital spaces, or what geographers have often referred to as 'the virtual' (Shields 2003), 'virtual worlds' (Taylor 1997), 'virtual reality' (Hillis 1996), 'virtual place' (Adams 1998) or 'cyberspace' (Dodge and Kitchen 2001) arguing that it is important to address the materiality of the virtual. Crang et al. (1999), in their collection of essays on 'virtual geographies', explore how new communication technologies are being used to produce new geographies and types of space (Adams 2009; Adams and Ghose 2003) and new 'networked affects' (Hillis, Paasonen and Petit 2015). Kitchin and Dodge (2011) address what they term 'code/space' in order to understand more about how computer code is threaded through the fabric of people's everyday lives. They mount a convincing case that the production of space is increasingly dependent on code, and that code is written to produce space. Some examples of code/space include airport check-in areas, and networked offices and cafés that are transformed into workspaces by laptops and wireless access. James Ash (2010) offers insights on how videogame designers manipulate space in order to create particular affects (also see Anderson 2006 on affect and Ash 2013 on affective atmospheres and technology). David Morley (2002) explains how new technologies, by transgressing symbolic boundaries, have changed traditional ideas about private households and nation states.

These examples of research produced over the past decade illustrate that geographers and other social scientists have begun to think deeply about the ways in which space is being reconfigured by social media, code, 'cultures of internet' (Shields 1996), technologies of touch (Cranny-Francis 2013; Paterson 2006, 2007), machines (Mackenzie 2002, Thrift 2003), digital sensations (Hillis 1999), mobile phones (Madianou and Miller 2011; Rakow and Navarro 1993; Rubinstein, Makov and Sarel 2013) and software (Thrift and French 2002). Space and bodies are continually being reconfigured by various technologies and ways of communicating. Skype is no different in that it is affecting how we engage with people, our relationships, perspectives on embodiment and what bodies and objects can do, in short, their orientations (Ahmed 2006). Skype, and other similar synchronous audio-visual communication technologies, are not just reframing or mediating more intensely existing social relations in unchanged spaces. All social and spatial relations are already mediated (Horst and Miller 2012). While the world (sociality and spatiality) may appear natural it is, as Butler (1990) highlights, a series of performances which are reiterated to take on the appearance of the natural (also see Gregson and Rose 2000).

Finally then, Skype may be enabling conversations and generating a range of opportunities for particular kinds of activities that did not exist in the past but how exactly? How and in what ways are these interactions taking place? Many people (but certainly not all) have at least some experience of using Skype (usually at home but sometimes at work) but we are not necessarily aware of how and for what purposes others might also be using this platform – perhaps in similar or dissimilar ways from ourselves. As mentioned previously, this is a point in time when Skype seems both familiar and strange to many of us depending on a variety of factors, which makes it an excellent time to pause and consider.

Notes

1 For more information on Ahmed's life and works see Wikipedia, Sara Ahmed (2016).

2 Pausé, Wykes and Murray (2014) in their edited collection *Queering Fat Embodiment* also drawn on queer theory to inform thinking on body size and shape. The editors argue that while queer theory has long influenced fat studies, theirs is the first collection of work focused specifically on the critical and political potential of queering fat embodiment. The contributions to their book do not just draw parallels between fat and queer experiences but rather examine the intersection of fat and queer, pointing 'to the ways that heteronormativity operates as a regulatory apparatus which underwrites and governs the discourse on – and management of – the fat body' (Wykes 2014, 4).

3 For a useful definition of 'cyberspace' see Bell (2009, 468) who refers to it as 'space between the screens'. He also points out that 'A keen geographical eye will immediately notice the suffix '-space' that makes up half this new word ... [and that] this suffix alerts us to the spatiality, or geography, of cyberspace' (Bell 2009, 468) (also see Bell and Kennedy 2007).

4 Also see Mackenzie (2002) on how technology becomes part of living bodies and Paterson (2007) for a discussion of the role of touch across a range of experiences including aesthetics, digital design and visual impairment and touch therapies.

3 Interviewing

Face-to-face and on Skype

Over the past decade social scientists, including geographers, have begun to collect data via the internet (Batinic, Reips and Bosnjak 2002; Hanna 2012; Jones 1999; Kinsley 2013b). These web-based methods are numerous and wide-ranging including online questionnaires, cyber-ethnographies, online content-analysis and online interviews[1] and online focus groups just to name a few. There is little doubt that the transformation in how people encounter each other (Baym 2010) such as sending emails, texts, instant messaging, photo-sharing, and playing online games using a range of devices including computers, iPads and mobile phones has extended to how research is conducted. In the emerging literature on online methods attention is being paid to the many pragmatic issues involved, for example, design, building rapport and ethical considerations (Chen and Hinton 1999; Madge 2010; Madge and O'Connor 2002; O'Connor et al. 2008) but few are focusing on how using the different mediums might *feel* for researchers and participants (but see Pile 2011 on the unconscious dimensions of the rapport between researcher and researched).

Interestingly, running alongside this trend towards online methods is another trend, and that is to use more performative methodologies. By performative methodologies I mean methods that focus on how different bodily practices involving the multiple senses of sound, smell, taste, touch and vision can be used to produce meaning and data. For example, Longhurst, Ho and Johnston (2008) use 'the body' as 'instrument of research' in a research project which involved cooking and eating with migrant women, Colls (2006) goes shopping with and tries on clothes with large women, and Waitt and Duffy (2010) listen to music at an outdoor festival.

It is perhaps somewhat ironic that both these trends – online methodologies and performative methodologies – are unfolding at the same time. In short, there has been simultaneous uptake of online methodologies that have no physical contact with people and of performative methodologies that focus more critically on bodily performance, taking seriously non-verbal interactions as a way of capturing meaning that goes beyond words. In carrying out this research I was cognoscente of these two methodological developments (online and performative) and deeply interested in what difference the body might make in relation to having an online or offline presence.

In this chapter I reflect on this but first explain in detail the methodological process used to conduct the research. Ahmed (2006), as part of her queering phenomenology project, ponders what is at stake in looking behind, in bringing the background to the fore. Often in books the methodological process remains in the background, unarticulated but in this project I put it in the spotlight. The primary method engaged, talking with participants face-to-face and through Skype, was integral to the findings. Data were collected specifically for this book project and were derived from interviews with 39 participants – 25 women and 14 men. Participants were asked a series of questions around the themes of how Skype helps shape various relationships with family and friends, the spaces and places of Skype, using Skype for special occasions, using Skype for work, whether people thought it worked better or worse for some interactions (e.g. intimacy) than others, the visual dimensions of Skype, grandparents' use of Skype and technical issues with using Skype.

Because I am interested in the difference between bodies having an online and offline presence some interviews (25) were conducted face to face, usually in homes but occasionally in offices and the remaining (14) were interviewed via Skype. Thirteen of these 14 were with video, and one was with the researcher's video turned on but the participant's video turned off. In some instances where it was not possible to meet in person interviews had to be conducted via Skype. In other instances participants were offered the choice. The majority opted to meet in person. All participants were asked to reflect on how the interview might have felt different if they had been interviewed face to face or in person (in other words, the opposite of the medium used). This means that in this chapter not only is a description of the research project offered but also critical reflections on interviewing participants via Skype.

The participants

Before discussing interactions with research participants, however, it is important to know something about them as a group. At the beginning of each interview they were all asked a set of key questions. The answers indicate that the age range of participants stretched from late teens to 65+. The average age of participants was 44. It may be that a different story would have been told if the cohort were on average much younger or much older. There was no special experience that qualified people to be involved in this study except to be able to answer yes to the questions, 'have you used Skype before?' and 'would you be willing to talk about this to a researcher?' I conducted approximately one third of the interviews and two research participants conducted the remaining interviews. Each of us used our networks of colleagues, friends and family members to secure participants. As a result 13 of the 39 participants have some connection or other with tertiary education, for example, as a university programme co-ordinator, lecturer or tutor. Other participants when asked to identity their occupation listed government worker, psychologist, digital artist, medical receptionist, miner, dispatcher, receptionist, administration assistant, music teacher, property advisor, operations assistant, homemaker, primary school teacher, care worker and unemployed.

Participants were also asked who they live with and the arrangements varied considerably including people who live alone, with their husband, wife, partner, children, flat-or room-mate(s) and cat. In order to build up a picture of the group involved participants were also asked to identify their ethnicity. Again, this varied with people identifying amongst a variety of other categories as Pākehā,[2] Cook Island/Māori, Australian European, Māori, American, Thai, White British and Melanesian.

By way of contextual information all of the participants were asked how long they had been using Skype for. The newest user had been using it for just 2 months. The longest user has been using it since it first began 13 years ago and prior to that she had used earlier audio-visual platforms including iVisit. On average the 39 participants had been using Skype for 4.5 years. Sometimes it was difficult for people to remember when they first started using it and how long ago exactly but most were able to work it out by recalling particular events or interactions. People were also asked which devices they used for Skyping. A total of 23 mentioned laptop computers but there was also mention of desktop computers, iPads, tablets, mobile phones, television and video conference systems in workplaces. Finally, people were asked who they contact using Skype. As might be imagined the list was long and included an array of family members, friends, partners, clients, customers, students and colleagues both in New Zealand and overseas. Given how many people currently use Skype and that I did not have many parameters around the project it would have been possible to keep talking with many more participants. A decision, however, was taken to stop when I felt that interviews were not turning up much by way of new material. In the final instance, a cross-section of people were involved in the project by way of age, occupation, gender, years of usage, and purpose of use. Because the research was conducted using snowball sampling, many of the participants lived in Hamilton but not all. Some had previously lived in the city but moved away. The research assistants and I contacted people we thought might be engaging and have a story or two to tell us about using Skype. A few others heard about the research being conducted and signalled that they would be willing to be involved.

After every interview whether face-to-face or through Skype, the researcher (whether it be me or the research assistant) wrote a couple of paragraphs on 'impressions' of the interview, that is, feelings we were left with afterwards, anything unusual or unexpected that happened, thoughts about the conversation, the place in which the interview was conducted and challenges posed by the technology (for the Skype interviews). These notes were vital in providing some context through which to understand the interviews, particularly those that I did not conduct myself. I listened to recordings of all the interviews but the research assistant's impressions added yet another dimension.

While this project is not auto-biographical (Moss 2001) or auto-ethnographic it does rely at least somewhat on my own experiences of using Skype. Over the past decade I have with my family members engaged with various digital media and adopted a range of technologies. Various cell phones, smartphones, desktop computers, laptops, Xbox and Xbox Live, GameBoys, PlayStations, iPods, iPads and digital cameras have come and gone from my life at various times. Using the desktop computer, laptop, iPad and smartphone to Skype with video has been a more recent experience and a significant one. I Skype (with video), and occasionally use Facetime, with various people as part of my home and work life. I am not a daily user but usually in the space of a week I have a couple of calls. I have documented some of these calls over the period prior to and during the writing of this book. For example, Skyping my adult son who lives in another city in Aotearoa New Zealand and seeing him 'at home' has helped me think through some of the issues, and in part motivated the project. My own mother, aged in her eighties, refuses to engage with digital media of any sort and does not own a computer. She, however, enjoys talking and seeing her other daughter and grandchildren on Skype when I make calls. I realize that over the years my own behaviours in relation to Skype have changed, for example, I am now less self-conscious about how I look. Some of these experiences are threaded through the book.

Feeling the interviews

In this section I attempt to tease out what might be some of the differences between face-to-face interviews and Skype interviews – that is, how they might *feel* different mainly for participants rather than researchers. The aim is not to offer a comparison between online and offline methodologies, ultimately preferring one over the other, but instead to tease out some of the emotional and affectual dimensions of the two different mediums. Research that compares online and offline methodologies has already been conducted. Hannah Deakin and Kelly Wakefield (2013) argue that for them Skype was a 'favoured choice' over in person interviews because it offered geographical proximity. Others (O'Connor et al. 2008; Opdenakker 2006) prefer in person interviews because they provide non-verbal cues.

This is perhaps not surprising given that as Miller and Sinanan (2014, 5, italics in original) suggest, it is easy to drop into discourses that 'lament[s] for a passing world of *real* relationships based on true social life that are most fully established by face-to-face communication' when discussing new digital technologies. As discussed earlier, however, all communications whether they be via a screen or in person, are in some way mediated. Simply judging online against offline or in person against Skype interviewing may unwittingly hinder closer questioning of how the two activities and spaces are mutually constituted and how emotion and affect flow across the bodies of researchers and participants in both real and virtual space. This alternative approach of attempting to ascertain how interviews online and offline *feel* may be of interest to other researchers in a range of disciplinary areas who are also currently using Skype and other similar platforms to conduct interviews or focus groups.

In order to ascertain how emotion and affect moves through bodies face-to-face and via digital media, towards the end of each interview participants were asked how they felt our interview might have been similar or different if it had been conducted via Skype or in person. In the case of Francine, who was interviewed via Skype, this prompted the answer: 'At least in person there would have been a cup of tea' (see Adams-Hutcheson and Longhurst, under consideration). The remainder of this chapter presents participants' reflections on this question.

First though, from my perspective, the interviews (individual, couple and group) which were conducted face-to-face provided an opportunity to *feel* the conversation in a different kind of way than the interviews conducted through Skype. Face-to-face interviews, for me, created an intersubjective, shared space in which thoughts and feelings were able to oscillate (Bondi 2003, 2005; Longhurst, Ho and Johnston 2008). As Gail Hutcheson (2013) notes, the 'atmosphere' matters. It is important to pause to 'contemplate the unconscious and embodied nature of conversational flows within research ... the way in which bodies can *impress* on each other' (Hutcheson 2013, 478, italics in original). Bodily gestures such as a sideway glance and actions such as keeping a cell phone close during the interview, making a cup of tea or pointing to an object in the room were obvious and oftentimes signalled emotion and affect flowing across bodies in the research setting. My physical body in the room was the primary filter through which emotions and interactions in the interviews flowed. It was my 'instrument of research' (Longhurst et al. 2008, 208).

These reflections risk sounding like a reiteration of the trope that somehow face-to-face is less mediated than through Skype, more natural as it were. But, to be clear, it is more that when I was face-to-face with participants all of my senses were engaged – smell, taste (when we shared food and beverages), sight, touch and audio. This did not make the research encounters better or worse, more or less natural, but it did contribute towards the way they *felt*, to the way emotion and affect flowed across the space.

When I interviewed people using Skype different senses were engaged – vision and aural but not touch, smell and taste. I could see and hear participants but not touch them, or share the same aromas or taste the same food as them. These are not senses that most of us tend to engage overly in research interviews anyway and yet without them to regulate the space the interactions did *feel* different.[3] There is often an ebb and flow to conversations, including research conversations or interviews. Sometimes this is achieved in interviews and small focus groups, sometimes it is not. It can be facilitated by sharing food or a beverage or by a handshake. This does not provide a guarantee and nor are these they only ways in which people can connect but they are often significant.

I was interested in hearing from participants what they thought about being interviewed face-to-face or through Skype – how it felt for them, their preference, whether it might have been different if the interview had *not* been face-to-face/via Skype, and whether sharing food and drinks during an interview was important. Given the wide range of participants, as might be expected, there were significant differences in the responses to this line of questioning and these responses did not correlate directly with particular aspects of subjectivity such as age, social class, or ethnicity. It was not simply a case of younger people feeling comfortable being interviewed through Skype and older people preferring face-to-face.

Four participants reported feeling that the flow of conversation in research interviews was *not* dependent on whether it was via Skype or in person. Either way the conversation could be relaxed or feel uncomfortably punctuated (sometimes both). Colin[4] aged in his sixties, for example, points out: 'Face to face interview is sometimes difficult; some people manage them very well and other people don't ... Some people perform brilliantly in lots of situations [e.g. Skype] with communication and other people don't.'[5] For Colin it is not so much about the medium being used but how proficient people are at communicating and putting others at ease. Nigel, aged in his early twenties, who Skypes friends and family overseas, was also 'neutral' about being interviewed in person or via Skype saying: 'It's a bit of a moot point for me. I wouldn't have minded if it was over Skype.' Beverley too claims: 'Honestly, either would have been fine for me. I think this is the kind of interaction that it's absolutely fine to have over Skype or in person, precisely this kind of thing where it's primarily about the conversation. So in that context it's fine'. Brianna, aged in her early twenties, when asked how she felt about being interviewed via Skype replied: 'It's cool.'

While these participants reported feeling equally comfortable with being interviewed either online or offline the majority of the 39 participants felt Skype was more likely to result in a sense of discomfort. Hunter, a university professor, says: 'Talking to a screen rather than to a person, you do feel a kind of distancing, a degree of distancing ... Lack of intimacy, lack of body language. I think it's an important form of communication but I'd hate to think it becomes the only form of communication.' Anong explains when asked how she feels about being interviewed in person: 'Initially being in the interview is awkward but with the person to person it kind of eases things gradually. For me in person is better.' In person, Anong and the interviewer were able to establish a rapport that Anong felt might have been more difficult to establish via Skype. Bodily presence in this instance was important. The interview took place at the researcher's dining room table. Sharing food and conversation Anong appeared to grow more confident in her answers as the evening progressed. Howard also preferred to talk in person and when asked if he would have preferred to have been interviewed via Skype he replied: 'No ... I love this.'

Harold and Kathryn were also interviewed in person, together, and again like the aforementioned participants, this was their preference. Kathryn explains: 'With Skype there's a voice thing, there's a voice delay. You can't have a spontaneous conversation where I talk and you talk, we'd cancel each other out. You have to stop and wait for the other person to talk. So that's the thing with interviews on Skype versus face-to-face.' For Kathryn the technology created a disruption in the flow of the conversation that she did not like. Similar sentiments were expressed by Margo, who says being interviewed in person feels very 'fluent. I mean we've sat here, it's all very relaxed and we're just talking now and chatting.' Encompassed in the notion of the flow of conversation is a sense of comfort. At the interview with Margo and her husband James we each had a glass of wine. James rubbed his foot gently on the belly of the old dog who lay at his feet. The rhythms of daily life (and conversation) became 'part of the way things are', which can lead to an 'ontological predictability and security' (Edensor 2010, 8). It is not surprising therefore that Brittney comments: 'It's less awkward doing it in person. Via Skype I think I would be like, "um, um, ah, yeah, probably" something like that. It's more comfortable; it's more relaxing and easier to talk [in person].'

While the participants discussed thus far were able to articulate how they felt in relation to being interviewed via Skype or in person, interestingly others seemed unable to answer the seemingly simple question: 'How might this interview have been similar or different if it had been in person/via Skype?' Geron, for example, replies: 'In person is better' but when asked if he could explain *why* he said: 'Always better. I don't know [why] actually.' Desiree also found it difficult to answer saying: 'I'd rather do this [in person] than Skype or something. If I have choice I'd rather be in person.'

Describing exactly how interactions feel face-to-face rather than via a screen, that is, how an image rather than a real body might reconfigure a relationship between touch and vision and the other senses and our spatial awareness can be difficult to articulate. Research interviews are not exactly a daily interaction that we are used to and yet most (but certainly not all) people still probably have a sense of how they expect them to unfold – interviewer/participant arrives, introductions are made, we get comfortable with one another, there are questions and answers, the conversation flows, people are thanked and farewelled. This usually happens as we share the same place, sounds, smells, food, drinks and so on. Things unfold differently when these interactions happen through screens. The next section discusses in more detail how and why some bodies may be confounded or disorientated when interviews take place using Skype. The body's sensory capacities change from what we are used to.

Shifting senses

Bodily closeness and the desire to come into contact skin to skin is part and parcel of social relations. Although often thought of as mundane, shaking hands or a touch on the arm, shoulder or elbow, for example, can be integral to establishing a rapport. The aim here is not to simply separate out touch from other senses but to acknowledge that it does provide an important and often underutilized framework for exploring emotional and affective relationships encompassed within interviewing spaces. Haptic geographies bring into the foreground bodily boundaries, the fleshy body and how various forms of touching are related to researching as a social process. Since Paul Rodaway first addressed haptic geographies in 1994 it has been recognized that formulations of touch go well beyond the immediate level of the skin – cutaneous touch (Dixon and Straughan 2010; Johnston 2012; Morrison 2012; Paterson 2009). Haptic geographies include bodily sensations, encounters and performances that aim to capture how 'bodies stare at each other' (Slocum 2008, 854). Touch is incorporated in a number of ways including bending the idea of touch beyond bodies to include how Skype is a device for keeping 'in touch'. It enables people to keep in touch, and to experience touching moments but *not* to physically touch the person or things that appear on the screen (Ash 2009).

Beverly usefully captures this saying: 'I guess it's [Skype] best for regular but low emotional intensity stuff; staying in touch … [but] it's really no substitute in situations where someone is genuinely suffering or upset because in those situations it's not really talk that people need; you need to be able to cook them dinner one night or you need to be able to sit with them and drink a beer in silence'. In these cases touch or at least physical bodily proximity is important. When Margo and James were asked how they felt about being interviewed in person and how it might have been a different experience from being interviewed via Skype, James reflects:

> People experience places differently but if you're in the same space you'd have your shared experiences based on the fact that you're humans in the same space … people will experience that differently but you're gonna have more of a chance of having shared experiences if you share it. … It's about constructive reality where we've got a shared construct, whereas you do have a shared element, but your reality is different. You're in different heat, different locations, different spaces.

Eva says that she thought about doing the interview by Skype but decided that she did not want to because 'I do find it a divider … it's not for me, a world you can walk into. It's another world.' It is a world where rhythms, viscerality and the senses are reconfigured. In years to come many of us may understand the variegated nuances and feel of online spaces more intimately but at the moment there is still often a sense that it is 'another world'. This world, for Eva, who had described herself earlier as quite a 'touchy feely person', does not always feel comfortable using Skype. It is a space in which it is not possible to share 'lovely baking', as discussed by Marie, who said if she had been interviewed in person she would have baked for the occasion.

Sharing food and drinks is one of the activities people often use to create an ambience of warmth and friendly interaction in research interviews (Bain and Nash 2006; Longhurst et al. 2008, 2009). The smell of baking, whistle of a kettle, the clinking of glasses, the shaking of hands and so on build a complex and multisensory picture of research that goes well beyond 'talk'. Haptic geographies are shaped by the multisensory complexity of touch and the other senses working together to produce haptic experiences (Morrison 2012; Paterson 2009) and knowledge creation in the research settings.

Having said this, the aim in this chapter has not been to simply replicate a binary where face-to-face interviewing is heralded as the *only*, most effective, and most unmediated way to facilitate tactile and visceral interview experiences. As with the complexities inherent in theories of emotion and affect, online and offline is far more multifarious and entangled than just bodily presence or absence. There are instances where Skype opens up possibilities for different kinds of intimacy and touch. Miller and Sinanan (2014) note that people sometimes touch the computer screen, enjoy the tactility of a device such as a cell phone in their hand (see also Longhurst 2013). The materiality of the organic (bodies) and inorganic (devices used for digital media) comingle. Kinsley (2013a, 364) makes the point: 'materiality is not a passive collection of stuff that is distinct from organic life'. Together the organic and inorganic prompt particular emotions and affects. Skype can sometimes allow conversations to progress in ways that feel comfortable and appropriate, while it is the physically co-present where things seem forced and unreal (Miller and Sinanan 2014).

Some participants, such as Izzie, preferred to use Skype for the interview. She says: 'it is quite convenient having it in my own house [laughter]. It's nice to just do it and not have to go anywhere to be honest ... it is nice yeah. I'm happy being interviewed this way.' Kieran, aged in his late teens, also preferred it but not because it enabled him to stay home but because he felt it offered him the opportunity to stop the interview if he so chose thus enabling him to relax. Kieran was asked whether, if he had done the interview in person at his university halls of residence rather than using Skype, it would have been any different. He responds: 'Yeah, I probably actually would be less relaxed I imagine. It's more relaxing for me just because you're not [voice trails off], I don't know, I can just quit at any time and run away or something.' The interviewer followed up with the questions: 'So, sometimes the more emotional, physical distance that Skype creates can be a useful thing too? You don't always want to be up close and personal?' to which Kieran responds: 'Yeah. Especially with having to talk to scary bosses or something, I would prefer to probably do it over Skype.'

Quentin agrees with Kieran that sometimes being interviewed in person can feel too confronting. He says that while he feels being interviewed via Skype was appropriate in this instance 'given the subject matter' [Skype], he continues: 'But if I had been divulging more kind of personal information or stuff that was emotionally charged, it would be way better via Skype versus in person.' Travis, aged in his sixties, also made the point that doing the interview via synchronous audio-visual link (in his case using Facetime) was appropriate given the research subject. Olivia was interviewed in person, which she enjoyed because she knew the interviewer but she said if she did not she 'would probably choose to do it on Skype ... it just gives you that little bit of, not anonymity but ... distance'. Again, Skype here is seen as useful in being able create a space that feels comfortable or orientated, sometimes through offering distance. There is not necessarily an easy direct correlation between offline space, real bodies, and comfort.

Clearly participant and researcher rapport is a delicate balancing act, whether it be online or offline, of being close but not too close, remaining professional but relating personally, and often with empathy (Bondi 2005; see also Pile 2010b, 2011). Gaining the appropriate intensity of rapport is not always easy since 'places are always in the process of becoming, seething with emergent qualities' (Edensor 2010, 3). They are, however, often made comfortable by people understanding the space they are inhabiting. It feels like things are familiar, lined up, clearly in perspective. If the interview is at home this might be through sharing food, settling or holding a baby, petting animals and talking with other household members. If the interview is in a workplace it might be by colleagues dropping in to exchange information or sharing a cup of coffee. If it is in a café it might be about customers coming and going, food and drinks being served, conversation and music. Many of us are less comfortable with how to 'perform' on Skype (Miller and Sinanan 2014). Currently, for example, we do not tend to leave Skype on in the background for long periods as we go about household activities but instead use it more like an earlier technology – the telephone – to communicate. As is discussed in Chapter 4 people are often unsure about seeing themselves in the side frame, which bit of their bodies to display, how close up to be and what is going on in the background.

Although interviews can also be awkward in a face-to-face situation and relaxed conversation is not always guaranteed, in Skype interviews there are limited opportunities to utilize other sensorial modes of engagement to smooth interactions. Sometimes smoothing interactions is made even more difficult by the potential for Skype to not function as expected, resulting in calls not transmitting, connections dropping, sound quality being poor and images pixelating or freezing. Not quite knowing how to *be* (perform) on Skype is worthy of consideration but so too is the potential disruption to our expected perception prompted by technical problems.

Katie explains:

> When you emailed me I was like, oh, I hate Skype but I'll do the interview here [via Skype]. I think I would use it more if the technology was better. It's not practical. Today is fantastic but this is the only good Skype conversation I've probably ever had. Every time it's interference. It's, 'Are you there?' 'Can you hear me?' They're stalling. They're pixelated. If that wasn't the case I'd be willing to Skype a lot more.

Manee feels that in person was better because with Skype there is always the potential for the technology to not work. She explains:

> If Skype works then it would've been okay but I think there would probably be some level of anxiety beforehand … I might have felt that

maybe it's too much trouble. Whereas if you just sit down and start talking, that's less of a hassle than having to go to a computer and making sure that it works.

To conclude this section, many of the responses from participants when asked how they felt about the medium through which they were interviewed are similar to those provided in relation to the themes of friends, family, home, work, love and romance. While some participants felt orientated and comfortable using Skype for a research interview others did not. This research reveals that people now often feel quite comfortable using Skype to connect with friends, family and loved ones but less so with colleagues and others they do not know. The reflective responses about the interview therefore are perhaps to be expected. Some participants were known to the interviewer, others were not. Clearly many factors are at play here but it is evident that, as Ash (2009, 2121) argues: 'the same material technologies can produce different phenomenal experiences'. While interviews (individual, couples and a group of three) were the main source of data they were not the only source of data. In the section that follows is a discussion of others sources also used.

Internet sources or 'vulgar geographies'

Kinsley (2016) says that he rather provocatively titles his article 'Vulgar geographies' to make the point that cultural geographers need to be effective at accommodating popular culture in their work. There is, he explains, 'still a paucity of work in geography on "low", "mainstream", "popular" or "vernacular" cultural geography'. (Kinsley 2016, no page number yet). Material about and contained within the ordinary spaces of life was incredibly valuable for the purposes of this research.

Although there is a rich literature on digital media and technology emerging in social and cultural geography, women and gender studies, digital humanities and a variety of other disciplines on many occasions I still found I was searching the internet for news stories, blogs, Facebook posts and Tweets on the various topics that I was writing about. As lecturers we often warn our students to be cautious of such sources – information is not verified, ideas are not peer-reviewed, anonymity can disinhibit people from making hateful comments online and 'Wikipedia is not the one and only source of knowledge' but the bottom line is that there is a wealth of important information online that academics can usefully draw on. By way of example, using the Google search engine I entered phrases such as 'distance relationships', 'Skype sex' (see Baker and Whitty 2008 on researching romance and sexuality online), 'mothers using Skype', 'grandparents using Skype' and 'Skype at weddings' to generate information to inform specific chapters. I followed threads of discussion that appeared to be relevant on topics such as family; kids; Skype moments; birthdays; user stories; Skype interviews: tips and tricks; screen sharing; stay together; Skype in the classroom; and Skype at work. I also searched blogs about Skype. For example, in 2011 Skype, to celebrate their 11th 'birthday' invited people to blog using the hashtag *Your #ILOVESKYPE Stories* (Caukin 2011). People contributed a range of stories about the purposes for which they have used Skype. Material such as this was really useful.

So too were posts to Reddit, the self-proclaimed front page of the internet. Using Reddit, I searched phrases such as 'Grandparents using skype' and 'Skype at weddings'. I opened each link or if there were too many posts to open I chose threads that looked relevant to themes I had been thinking about such as space, place, identity, distance, embodiment, gender, performativity, touch, distance, visuality and bodies.

Reddit proved particularly useful for Chapter 7 'Skype Sex' since participants were understandably somewhat reticent to discuss this, or had little experience. The 39 participants were asked whether they would consider, had in the past, or currently use Skype for sexual and romantic encounters but this question did not generate much discussion. Sometimes I used the term 'intimate' rather than 'sexual' because I sensed participants might not be keen to discuss their experiences. Therefore, I needed to find another way of pursuing this topic. Reddit helped provide some context for participants' comments about their sexual experiences (or lack of) on Skype. I searched the phrase 'Skype sex'. I did not pay attention to all the posts since more than 22,000 entries appear under the heading 'Skype sex'. Sub-headings are wide-ranging, for example, dirty penpals, dating advice, confessions, and DeadBedrooms. I concentrated on posts that appeared relevant to issues raised by participants such as 'Relationships' and 'Long Distance' and more likely to contain threads about sexual and romantic relationships often considered more mundane between partners, lovers, husbands and wives.

Kinsley (2016) suggests that cultural geography would benefit from attending more to 'the popular', that is, to 'everyday' cultural forms. This would assist geographers in helping make sense of the world. In the 1980s the 'cultural turn' drew heavily from literary theory to inform ways of thinking and knowing. Several decades later, we need to ask 'What does it mean to think in terms of video, audio, the smart phone and the tablet?' (Kinsley 2016, no page number).

At times throughout the chapters I draw on 'vulgar geographies', that is, information collected on and about video, audio, smart phones and tablets alongside academic texts and interview data. These sources from popular culture have helped me think through the experiences of others, beyond the participants, supplementing their narratives. Increasingly social scientists are turning to online sources since the interactions they represent, whether they be chat sessions, Facebook posts, Tweets, emails or blogs, are arguably just as important as offline or real life interactions (see Driscoll and Greg 2010 on virtual ethnography).

Researchers often use their bodies as 'instruments of research' (Longhurst et al. 2008) but in this chapter I have attempted to extend thinking on how exactly this is the case when the researcher's body and the bodies of partici-pants are filtered through screens. Being aware of one's own embodiment as a researcher, and the bodies of those who are being interviewed, means acknowledging a range of emotions and affects that take place in the research setting whether this be online or offline (Gorton 2007).

Arguing that emotions and affects flow or 'fold' (Ash 2015a) through all spaces – online and offline – does not imply that space itself does not matter. It does. Nevertheless I do not want to advocate one medium – online or offline – over another. In this research most participants (but certainly not all) felt more at ease with real body proximity than they did being interviewed via Skype but this may change in the future as people become increasingly used to screens as an interface to connection. To understand participants' comments in more depth attention was paid to touch in order to highlight the mundane nuances of everyday research practice. Being able to touch someone on the shoulder, share food and drink, absorb the atmosphere in the room and be part of the comings and goings of life taking place in and around interviews enabled participants and the researcher to establish a rapport that tended for most (but certainly not all) to feel comfortable.

Often while interviewing there is far more going on than what initially might be considered. Interviews are performative. Many different bodies can be present, semi-present or co-present but remain invisible in research accounts (e.g. children listening in or needing attention, partners appearing in person or the background when using Skype, pets wanting attention). In some cases, for a time, the researcher becomes part of the household, café or workplace depending on where the interview is being held rather than an interruption to it. Therefore, bodily emotions, affects, sensations and tactile connections are co-joined to expose the reproduction of everyday life. Part of the atmosphere or ambience created by haptic moments is to keep the sociality of the space in play.

As I said at the beginning of this chapter, interestingly the focus on online methodologies as a tool for research has come at a time when the 'linguistic turn' in the social sciences has been challenged by methodological approaches that demand more attention be paid to non-linguistic elements (things that can be touched). There has been something of a turn towards practice and performance. Privileging discourse over matter and the materiality of things and people has been critiqued. The materiality of bodies, performance, movement, senses, emotion and affect needs to be taken seriously and to be grappled with *methodologically* (Longhurst and Johnston 2014; Sharp and Dowler 2011). This includes work on unconscious communication through psychoanalytic geographies (Bondi 2014, Hutcheson 2013, Kingsbury and Pile 2014) and how emotions and affects entangle researchers in particular ways 'in the power-saturated social structures of the field' (Lalibertè and Schurr 2015, 3). It also includes work on 'the interface envelope' (Ash 2015). Although James Ash focuses on gaming, his insights about how digital interfaces function to shape spatial and temporal perceptions can also, to some degree, be applied to relationships between people, including researchers and participants, using audio-visual synchronous platforms such as Skype.

The point of this chapter then was to illustrate that Skype is not just the subject of this research but also was an intimate part of its actual production. The aim, in keeping with Ahmed's (Ahmed 2006) project to queer phenomenology, was to look back at where and how the data came into being, to ensure that readers do not lose sight of it by simply looking forward to the research results or outcomes about people's use of Skype for specific purposes such as to see and talk with grandchildren, meet with a colleague, or have a sexual encounter with an absent partner. Before looking forward to address these themes in the next chapter though, I want to look – sideways.

Notes

1   See O'Connor et al. (2008) who argue that face-to-face and online interviews, synchronous and asynchronous, have their pros and cons, and that some are better suited to particular research questions and populations than others. Also, see Burns (2010) on email interviews.
2   The term Pākehā refers to Aotearoa New Zealand born people of white, European descent. Although the term Pākehā has been (and at times still is) highly contested in New Zealand (see Spoonley 1993) it is now used as a standard term of classification of ethnicity in the New Zealand Census. The term Māori is commonly used to refer to the tangata whenua (literally 'people of the land') or indigenous peoples in Aotearoa New Zealand. I use this term here but wish to problematize such use. As Spoonley (1993, xiii) points out, 'the word "Māori" is really a convenience for Pākehā to lump together divergent groups'. Māori often identify by iwi (tribe) rather than simply as Māori.
3   In Adams-Hutcheson and Longhurst (under consideration) we take the argument made here about bodily senses in a different direction by considering the role of *bodily rhythm* in constituting the social world. Lefebvre (2004, 15) argues that 'everywhere there is interaction between a place, a time and an expenditure of energy, there is *rhythm*' (emphasis in original; see also Edensor 2010). Drawing on 'rhythmanalysis' as an analytic lens we explore the specific practices of interviewing and the organization of social life within the interviews, offering the opportunity to rethink time-space as produced in the practice of research rather than as a container in which those practices are played out (Simpson 2012).
4   All participants in this study have been given pseudonyms to protect their anonymity.
5   Transcription is itself a form of analysis. It does not simply represent in some neutral fashion a conversation but rather constitutes a framework of understanding. My interest in analysing participants' stories was primarily with the content of discourse not in the moment-by-moment conversational coherence of the discussion. The transcription codes used throughout this book are as follows: '…' denotes omitted material; bolding denotes words or particles said with emphasis; comments in square brackets, such as [laughter], have been used to include non-verbal communication and events that help to give context to the conversation; and, commas, full stops, question marks and exclamation marks have been added in a manner designed to improve the readability of the extracts while conveying their sense, as heard, as effectively as possible.

4 Selves, others, objects and space

By looking sideways it is possible to see glimpses of selves, others, objects and space on the screen when using Skype. Many of these themes – seeing ourselves in the box in the corner of the screen, whether we decide to do our hair before connecting, what we think when we see others, whether we show someone around our space using the webcam, and what we might see by way of other people's backgrounds – while seemingly banal, actually matter. They play a major role in the framing of interactions and therefore it makes sense to canvas them before thinking more specifically about family, friends and loved ones' use of Skype (Chapter 5), using Skype at work (Chapter 6) and Skype sex, love and romance (Chapter 7).

The self in the box

Miller and Sinanan (2014) explain webcam is really the first time that people have been able to see who they are. They concede that we have in the past had mirrors, photographs, and been able to see ourselves on video or film but they argue that these tend to be relatively self-conscious appearances that we have created for a particular reason. Miller and Sinanan (2014, 24) continue:

> There is ... a good reason why so many people who use Skype become drawn, after a while, from the person who occupied most of the screen to the small box nestled often in one corner, within which they are confronted by an animated version of themselves. The self that they are drawn to is one they have never seen before.

This self in the box is the way we look to others. It intrigues us (Horst 2009). An image of the self is reflected back to the self. In real space we sometimes catch a glimpse of ourselves, for example, a reflection in a window, mirrored glass or a shiny surface, but when using Skype we are continually confronted with our own image. Natalie explains that when she Skypes her grandchildren who are a range of ages they are keen to see not only her, their grandmother, but also themselves: 'With the kids it's fantastic; you just want to see them and they want to see you, and actually they want to see themselves because in the little picture they see themselves and they look at it all the time.'

I remember vividly the first audio-visual conference I attended approximately 10 years ago. With a sense of horror at seeing my own image in a box in the top corner of the large screen I sat rigid for more than hour, taking care not to touch my nose, brush the hair off my face, or move too much – face or body – in any direction. After the event my muscles felt stiff as a result of my highly self-conscious performance on camera. Now, at similar events I hardly notice my image on the screen and am less concerned about bodily and facial gestures. Sometimes I leave my seat to open a window or welcome a colleague, drink a cup of coffee, or look at emails on my phone if it seems appropriate – something I would never have done until just a year or two ago. Getting to this point of feeling less self-conscious, unconscious even, can take time and not all participants in this research yet felt comfortable with their own image in the box in the corner of the screen when using Skype.

Sometimes the image of oneself can feel disorientating. It can also be difficult to know how to position oneself in the shot, not just one's head and shoulders but the rest of the body, and where exactly we should be in relation to others around us. Debbie, a university lecturer aged in her mid-forties, makes the following comments

DEBBIE: You know the little box on the right hand side, at least on my settings, where you see yourself, for me, it's terribly distracting. I should just turn it off but I have trained myself, I'm much better now, to look at the camera on the top so the little webcam camera in my laptop which is inbuilt. I've worked out just to focus on that and even though I can't see myself I know they can see my whole face ... this is something I learned after doing a job interview, a pre-interview job interview, on Skype. That was really useful actually because it made me realise how you're seen because if you're looking down.

INTERVIEWER: If you're looking at the image of your parents or your interviewee, interviewer, you're not actually looking at them?

DEBBIE: Yeah, that's right, but I find that really difficult 'cause if I'm looking at the camera and elevating my head just that little bit, I can't see them. If I'm looking down at the little box I'm just getting distracted by whether my hair looks ridiculous or whatever it might be.

INTERVIEWER: Before you know it, you think it's a mirror.

DEBBIE: Exactly. I'm not, I don't think, particularly a vain person but you can't help be captivated by your own image. Maybe I am.

INTERVIEWER: ... When you see your parents is it, it's a head and shoulders shot I presume 'cause they're sitting quite close to a table?

DEBBIE: Yes, 'cause they're sitting down, yeah, that's right.

INTERVIEWER: Do you ever think you'd like to see more of them on screen? Like getting up and active and walking around or more of their bodies?

DEBBIE: Yeah ... I feel like we see my parents reasonably regularly so they're in their seventies now and I get a sense of what's changed in-between but it would be interesting. I think their study is a bit of a confined space so for my Mum particularly who's not as physically able as my father, to get up and down might be a bit of a pain for her. Whereas I often, 'cause I sometimes feel like [partner's name] is off to one side so I'll make him come in, him and [daughter's name], I'll put them at the centre of the call after a while and I'll stand up behind. Then I hover and I sometimes go and potter in the kitchen.

I quote this conversation with Debbie in full because it addresses many of the everyday issues I have found myself pondering since embarking on this research. These include learning to see ourselves, to see others, where to look at the multiple boxes on a screen or sometimes multiple screens, whether to sit or stand, stay still or move, how many of us can fit in the frame, who should be at the front and who should be at the back, whether to do our hair and make-up, change our shirt, or be in bed or at the kitchen table. These are all things we must learn in 'doing' Skype but how exactly? New bodily performances have to be repeated (Butler 1990) to establish norms in relation to the 'doing'. Brittney first tried Skyping a few years ago but then 'refused' to continue because she 'hated' what she looked like in the box on the side. She says it does not bother her now (she has grown used to it) and in fact these days has come to think that because images on Skype are not always crystal clear it can be a useful device for hiding some of the 'imperfections' on her face that would otherwise be noticeable in offline space.

Participants in this study were asked many of the aforementioned questions about how they feel about seeing themselves in the box on the side, whether they tend to face the camera directly just revealing head and shoulders, whether they get up and move around, and whether there are any gender differences in relation to the attention paid to bodily performance when using Skype. The answers to these questions and a number of others were highly variable. Participants were divided approximately half and half as to whether they were concerned about their appearance when Skyping family or friends. Travis, who is aged in his mid-sixties and lives with his wife, explains

I'm using it with friends and it's just the way you are so it doesn't matter; friends and family they just take you as you are anyway. Quite often we've been still in our pyjamas or whatever or just in tee-shirts or whatever just sitting around home when we start talking to someone.

Having said this, though, Travis then adds that he does think that being seen makes people a bit more aware of their appearance and that he made sure he did his hair prior to our interview on Skype.

The difference gender makes

Quentin, a researcher aged in his early forties who was interviewed on Skype, thinks that having webcam makes a difference to just a phone call or Skype without video and that maybe women are more aware of this than men. He says:

> Yeah definitely [webcam makes a difference] but it doesn't stop me picking my nose absentmindedly though, just joking [laughter]. But it wouldn't stop me if I hadn't had a shave or whatever and the person needed to talk or that was the only time we were gonna get to talk. It wouldn't stop me logging on, it wouldn't stop me using video. But female friends of mine, they will say 'I just got up' … some, not all, will say 'oh nah I look like shit at the moment', or 'just let me put my eyebrows on' or 'let me put my face on'. But usually my response is, 'don't worry about it. It's just me, shit'. And sometimes they'll just go 'oh alright' and their hair might be a mess or whatever, or they might have been crying and eyes are puffy or whatever but yeah, generally the dynamic of our relationship means that I'm not hung up on that stuff. It certainly seems to be gendered. I guess guys are paying more attention to their appearance but me, even if I look like shit, nah [no] I don't think it would stop me having the video.

Katie, who is aged in her early twenties and has recently moved from Aotearoa New Zealand to Australia, reinforces Quentin's point saying that she is short-sighted and likes to wear her glasses on Skype not so that she can see the image of others more clearly on the screen but so that her eyes do not shift from side to side too much. She is conscious that they do this more if she is not wearing her glasses. Katie explains:

> I guess with men I like to wear my glasses and make sure I've got mascara on at the minimum and looking okay. I'm conscious of my glasses so my eyes appear normal and I'm conscious of my makeup so that my appearance is what they would normally expect of me, especially not for people I'm really close to like [name of close male friend] for example. If I was in a relationship I probably wouldn't care. But it's especially for [men], oh no, it's for women too, for men and women that I am friends with and we could be good friends. But I would want to present myself how they would expect to see me in real life; and in real life those people would never see me without makeup or without my hair looking cute. Whether it was up or down, I would make sure that it was styled appropriately and that I was wearing at least a nice top. Even if I had no pants on I would be wearing a nice tee shirt!

It is perhaps unsurprising that there are now numerous websites aimed mainly at women which are devoted to 'how to look cute on Skype' (StyleBistro 2013, also see Kang 2012 on gendered media).

Barbara, a medical receptionist and Henare, a miner, both aged in their fifties, were interviewed together using Skype. They were asked if they knew they were going to be on webcam if they felt like they had to look or be dressed a certain way. Each replies candidly

BARBARA: Yeah, definitely. I've got no bra on now but I've got a singlet on. I'm not gonna put the camera down past my chest 'cause you don't need to see my saggy boobs! Definitely. Just in what I'm wearing.
INTERVIEWER: Did you do your hair and stuff like that or had you already done that before?
BARBARA: Yeah. I'd make sure at least my hair's combed; not sticking up like Billy Idol [English rock musician with punk style hair].
INTERVIEWER: Have you ever had a Skype conversation where you feel you look terrible?
BARBARA: No, I don't think so I've never not combed my hair for Skype. No.
INTERVIEWER: Do you think though that women and men respond differently to that visual aspect of Skype? Do you think guys care as much?
BARBARA: I don't know. I don't think so. No. [Turning to her husband, Henare] What do you reckon; no?
HENARE: No.
BARBARA: No, he doesn't care.
HENARE: I wear my hairy chest outfit.
BARBARA: He sits there with no top on and hairy chest and he doesn't care!
INTERVIEWER: And, the face-to-face screen contact of Skype, do you ever find it intense or confronting?
BARBARA: I do sometimes and I think, 'Oh God, I look terrible'. But then I'm like, 'Oh, what the hell; don't worry about it'.
INTERVIEWER: So that's only in reaction to yourself, not to the other people?
BARBARA: No. Only in reaction to myself.

When asked if Barbara finds the face-to-face shots on the screen 'intense or confronting' it is her own image that she finds thus, not the images of others. Brianna, who is aged in her early twenties and has been using Skype for approximately four years, is also highly conscious of how she looks on the screen. She says

I always care [about my appearance on screen] because Skype is different to talking to someone in person. That's how I feel about it anyway. People are going to look at you for an hour or two hours and your image is huge on their computer screen. You do think about it.

She continues explaining that if someone Skypes and it is not planned she will not answer. If, though, she has arranged a Skype session with someone she will put on make make-up and a top that makes her 'look good' even if she has pyjama pants and slippers on the bottom. Brianna says that she thinks women are far more self-conscious about their appearance on Skype than men who she says 'don't notice' or 'care' but 'they will tell you you look crap if you look crap'. In short, gender relations around body concerns and expectations that construct women as always needing to be conscious of their appearance and to better themselves (Murnen and Seabrook 2012) which have long been played out offline are also evident online. Some women claimed that they do not worry too much about their appearance but at times this seemed like maybe they did not want to appear vain or overly obsessed with self-image because in other statements they made they clearly were aware of how they looked and did make varying degrees of effort to manage their appearance ahead of a video call. This somewhat contradictory position is evident in Alex's comments.

Alex was interviewed as part of a group of three friends aged in their early 20s. She has been using Skype for approximately two years. To begin with, when asked if she considers how she looks on Skype, Alex replies:

> Definitely. I always think about the angle because when you hold it [the laptop computer] from down under your face your face actually looks disgusting [laughter]. So you kind of have to hold it either right in front of your face or a little bit above. That's the thing I **always** think about, where I'm holding the computer.

A minute or two later in the conversation, however, Alex's perspective seems to change as she says:

> No [I don't worry about how I look], I mean purely because the people that I Skype are close friends and family. I don't really care; they've already seen me without makeup and my hair looking crazy. But if it was like a boy that I was dating then I might, possibly.

There was actually a great deal of difference amongst participants in relation to how they presented themselves on Skype, how they felt about seeing their image, whether they perceived there to be gender differences and how they talked about these things. It did not entirely map neatly onto gendered expectations although as suggested earlier there was some mapping. Katie says that she does not think there are gender differences and that it is more about individuals' insecurities.

> For example, with [name of a person she once worked with] we used to Skype quite a bit and one day he was on a computer and I randomly called him … He answered and we were talking for a while and I was like, 'Oh! I didn't know you wore reading glasses,' and they had a purple lens in them. For his visual impairment needed a coloured lens. And he was actually really embarrassed and he was like, 'Oh, sorry,' and just took them off. I'm like, 'No, fine. They look nice. I really like them.' And

he was like, 'Oh, no, I didn't want you to know that I wear these glasses but I have to wear them sometimes.' ... Yeah, I guess it probably just depends on the individual's insecurities. I think they more want to present themselves to people they know how those people would expect to see them in real life. I'd never seen [name of person] with those purple glasses on in real life and so he was embarrassed on Skype because he knew I wasn't expecting that.

Approximately three-quarters of the participants said that when Skyping family and good friends they did not tend to worry about their appearance. They might fix their hair, but women would typically not worry about make-up. Men and women said they would rather not be in pyjamas but again with people who are close they were not too concerned. Again though, this was certainly not the case for all. Queenie, a property advisor aged in her mid-sixties who enjoys Skyping her granddaughters in Australia says that she always makes an effort.

It is important because they [granddaughters] say to me, 'You've got orange on today Nana' so I make sure that I'm usually dressed well, and one night they said to me, 'Nana go and put your nightie on so we can see what colour you sleep in.' To them, especially to children, your appearance is really important. Yes I do, I make sure my hair is nice, I've got something decent on and I've got lipstick on.

As can be seen there were many and varied stories from participants about how they feel seeing themselves on screen and what they did, or did not do, to prepare for Skype with video calls. Factors such as age, gender, how long people had been using Skype, who they are calling, the purpose of the call, and whether they have a typical 'Kiwi' laid-back attitude[1] (e.g. Barbara with no bra and Henare with exposing his hairy chest) all made a difference to the bodily performance. Sometimes people might begin the call feeling highly aware of the small box to the side of the screen containing the self but after a while less self-conscious about it. Miller and Sinanan (2014, 25) note that for people seeing themselves on Skype:

There will almost inevitably be some aspects of that appearance which are unexpected or different from the previous imagination the viewer had of how they appear to others ... Even if there are no such discrepancies ... they are likely to have a heightened consciousness of themselves as occupying the gaze of others.

This was certainly the case for participants in this study. To return to Ahmed (2006, 1) in the Introduction to her book *Queer Phenomenology* which is titled 'Find Your Way' her opening question is: 'What does it mean to be orientated?' Many of us in relation to Skype are still finding our way, even after using it for several years. We are working out how to orientate ourselves, how to look, how to move, what and who to face, through the screen. Ash (2015a) says that life is increasingly becoming 'enveloped' as people become involved with more and more digital interfaces. Skype is a useful case in point. In the section that follows I move from thinking about the self and others on screen as in these new 'interface envelopes' (Ash 2015a) to also thinking about the spaces and places in which we Skype.

'Theatres of composition'

Dianna Fuss (2004, 1) writes: 'the theatre of composition is not an empty space but a place animated by the artefacts, mementos, machines, books, and furniture that frame any intellectual labour'. The phrase 'theatre of composition' feels especially apt for thinking about using Skype in that, as we have seen, people quite consciously compose or construct space in order to appear a certain way to others. Of course, for the purposes of this research, unlike Fuss I am not just talking about framing intellectual labour but life in general. Ahmed (2006), in discussing Husserl's work comments that he makes an impression on her when he offers a glimpse of the domesticity of his world. She is intrigued by, and imagines the objects around him – the writing table, the paper, a leather chair to one side but also the portions of the room that remain behind the philosopher's back, the parlour, the kitchen, the veranda that goes out to the garden where the children play in the summer-house ('yonder'), that is, the domestic spaces.

One of the things that interests me about Skype is that the spaces in which people are displayed are sometimes fixed (if they sit in one place and do not move much in front of the webcam) but other times mobile. If using Skype on a mobile device, or a laptop computer, then it is possible to see what is behind people's backs and to the sides and in front of them when they turn or move around. The 'here' and the 'yonder' can become one and the same, dramatically changing the perspective that most of us are used to. What is 'out the back' metaphorically and literally, can be revealed to the other person or people on the end of the cables. Some find this intriguing. Others find it utterly amazing or are 'blown away' by it, to use one of the participant's (Travis') terms.

For example, Fab, a lecturer in her late forties, says:

> I showed my parents what we've done in the garden, what trees we've planted. With the laptops and various tablets you can walk around the house and take a connection and show off the things, yeah, that's another thing. It doesn't necessarily have to do with Skype itself, it's more like the portability of devices these days and the iPad. I got my iPad and I can show my parents anything. I'm not a person who likes to show what they do. I don't post anything on Facebook. I don't show what I do in my life. I don't see any point in doing that. Same in Skype, apart from yes, the people, my parents, my sister, they're kind of curious about what our

house looks like, whatever around the area looks like and what the inside of the house looks like. We've got a new house so I showed them around the house with Skype.

The following chapter discusses children using Skype for 'show and tell' but clearly it is not just children who like to show others things and places using Skype.

Butler (1993, ix) explains that when she began writing her book on the materiality of bodies she found that '[n]ot only did bodies tend to indicate a world beyond themselves, but this movement beyond their own boundaries, a movement of the boundary itself, appeared to be quite central to what bodies "are"'. On Skype, when people show others around their space, the boundaries between bodies and environments sometimes appear to merge. Screens often display unexpected bits of bodies simultaneously with glimpses of rooms, gardens, streets and objects at unexpected angles as we move about pointing our devices in particular directions. Travis, who is retired, when asked about whether he shows people his surrounding environment says:

> Yeah, pick it up and show around the house or something like that; quite often done that when we've been away especially; when we've been overseas and talking to the kids on Face Time we'll flick it around and show where we are and show what's around us and they do the same as well sometimes ... Yeah it's very good; it just gives you an idea of where they are and perhaps what they're doing and a bit of atmosphere I guess. I enjoy it; I think it's good. I'm still blown away really by the technology that you can do it. It's good to see especially if they're talking about something and they can just pan around and show you what they're talking about; it's great.

Beverly, an Australian now living in Aotearoa New Zealand in her late 20s, on two occasions, found Skype useful for finding housemates. She explains that when she was living in England she spent a summer in the United States and so, to organize a sublet for that, she set up a Skype conversation with the person who she was going to be subletting from and her housemate who she would be sharing with. She then did the same thing in Aotearoa New Zealand.

> I put on the forms that I live in Hamilton, which I do; but I also have a room in a house in Auckland, which I organized before I came to New Zealand. I organized that through Skype as well. I had a Skype interview with one of the flatmates and then later with both of them ... In both cases we did a tour where they take their computer around and show it. It's really not that effective as a way of getting a sense of what a place is like. It's a little bit like, you know ... the distinction between maps and tours. It's like the extreme form of the tour where you just have no sense of where you are in relation to any of the other rooms. Even if you get a

sense of one particular room everything else is just like so jumbled up, that you arrive there and it's never anything like you saw; but it's enough to get a sense of the atmosphere of the place. The light and what it feels like to be in; whether it feels spacious enough, whether it feels pleasant. You have to be prepared, like all photographic things, I guess. The other main thing is to see whether you get on with them, whether you're gonna do that. I guess, as well, it kind of adds a sense... because organizing flatmates from a distance is always risky on both parts, and there is something about Skype where you get to see the people in the house that you're going to be living in; and vice versa, they can see you as a real person. I mean, obviously it's partly an illusion. Still it's a comforting illusion to be able to feel like there are real people on the other end of this internet ad.

As Beverly suggests, representations and the real are not the same – the affects created are different – but it is still possible to get an overall sense – a feel – of the people and place from Skype (see Karatzogianni and Kunstman 2012 on 'feelings, affect and technological change').

Exploring this idea Carsten Stage (2012) analysed user-generated videos of a Lady Gaga concert in Denmark on 20 October 2010 concluding that media and screens are technologies 'that under certain circumstances enable intensified relations between bodies, spaces, and significant others'. Media visually enriched the real-time experience of concert-goers. For a participant in this study, Manee, the screen did not so much enrich real-time experience but it did make her feel closer to her daughter who moved from Aotearoa New Zealand to China than if she had only been able to phone her.

When I was Skyping with my daughter in China before I actually went there, I asked her to show me the rooms and whatever, because I don't have any sense of what her accommodation looks like, what kind of apartment she's got. What does the kitchen look like, and what's in the lounge? That kind of thing. It was good to be able to see some of these things because then you can have some idea and not just completely imagine it. You could see this is what she's talking about, or when she's talking about whatever, then you can picture it because you've seen it on Skype already. And that enriches the communication.

Up until this point what has been under discussion is the deliberate display of objects, people and spaces but sometimes it is possible to glimpse in backgrounds things that are unintended. What is being referred to here are backgrounds in a spatial rather than a temporal sense. In discussing the term 'background' Ahmed (2006) reminds readers that it carries both meanings. For example, when we tell a story we might provide background information – what happened in the past – referring to background in a temporal sense. Here, in relation to Skype, however, I am referring to a spatial sense. Ahmed

(2006, 38) explains: 'A background can refer to the "ground or parts situated in the rear" (such as rooms in the back of the house), or to the portions of the picture represented at a distance, which in turn allows what is "in" the foreground to acquire the shape that it does, as a figure or object. Both of these meanings point to the "spatiality" of the background.'

When I asked Ewan, a government employee, whether he had ever seen anything in the background of a Skype call he muses:

> I can't remember seeing that but I have been aware of that with the boys [his two young adult sons], wondering; in some ways, hoping I wouldn't, 'cause I didn't want them to be in a position of feeling silly or anything like that; that I hadn't unintentionally crossed the line of privacy ... there's that sense of not wanting to crowd them unreasonably, and this sort of imposing myself upon them into the middle of their space when they may not be prepared. For a while I used to think about that in the back of my mind ... I feel that Skyping is more intrusive. I feel that on the phone they can keep doing something while they are talking. I feel a bit funny about calling on Skype, or Facetime ... I don't want to be imposing on them.

In this instance Ewan had not seen anything or anyone on Skype that was unintended but he has worried that he might and did not want this to be awkward with his sons who were both establishing adult lives in other cities.

There can be awkward moments when the private and the public collide. Once when Imogen, a digital artist aged in her late forties who is a New Zealander but now lives in a city in Europe, was with three other members of an art performance group and contacted a fourth member, the fourth member's boyfriend walked past naked in the background not realising he was on camera. The three members watching all burst out laughing and the one whose boyfriend it was turned to him and said 'Surprise!' In a similar vein, Olivia who occasionally has board meetings in the evening, Skyping from her home, says that she has to be careful that her fiancé does not make an appearance in his towel after having a shower. She explains: 'I have been nervous because he tends to walk around the house naked a lot ... Now I tell him, "[name of fiancé] I'm on a Skype meeting, you need to put some clothes on or just don't walk past me" ... Yeah, apart from that there's nothing I don't want anyone to know. If I'm having a board meeting at night I tend to keep my glass of wine out of sight.'

Olivia is keen during her Skype meetings to keep a couple of things in the closet (Brown 2000; Sedgwick 1990) – namely her semiclad fiancé and her glass of wine. As Michael Brown (2000, 1, italics in original) notes the closet is 'a patently geographical signifier ... a *spatial* metaphor: a way of talking about power'. It is about visibility or lack of (see Tucker 2009 on 'queer visibilities'). If Olivia makes too much of her domestic life visible it may undermine her professional credibility as a board member.

Still on the theme of visibility, Denise is a university lecturer from the United Kingdom who now lives in New Zealand with her husband and two children. When asked the question about not disclosing things to others on screen or seeing anything not intended for her when Skyping, she says that while this has not happened to her, it has been the case for her sister:

> It's harder for my sister 'cause she's got three kids so they're harder to control when the Skype's there, whereas my kids are older or not there so it's just me and [partner's name]. But with my sister the kids have either said something or done something that she can see and so my sister can't cover up anything. Or visibly, if the kids had something, a bandage or a bruise or something, my mother would immediately see them and want to ask. So you wouldn't be able to cover that up. But I know with my sister there's been a few things that my mother... Basically my mother can see everything that's going on and the kids just reveal everything as well. So for her more so but not so much for me.

Trying to manage the bodily performance of children can be challenging. Reading this scenario through Ahmed (2006) it seems that the object of the bandage was not intended to be part of the view for the subject (mother) on the screen and yet it appeared prompting a conversation that would otherwise not have taken place. The bandage was supposed to remain in the background but Skype had the effect of queering the space, changing the perspective, and bringing it from the fringe to the front, in essence, bringing it into view. Denise's sister could have removed the children from view entirely, not involving them in the Skype call (and thereby keeping the children's injuries secret or hidden from view) but as Sedgwick (1990) notes the closet is an 'open secret', a 'knowing by not knowing'. If Denise's sister had done this, therefore, her mother might have suspected that something was wrong.

To return to Olivia's story about having board meetings using Skype from her home in the evenings, she says that on one occasion when she saw someone else's home she found herself thinking:

> 'Oh that's what their house looks like,' or 'Oh that's what their [voice trails off]', you know [I've] done exactly what I don't want people to do to me, which is why I keep my background blank. So I have kind of just observed through Skype, 'Oh that is such and such'. Or not necessarily that they don't want me to know but maybe another board member's son pops his head past and 'Oh didn't realise her son was 12 years old, I thought he was only five.' Never anything [that] no one wants me to know but just kind of made some observations.

Again, what becomes apparent here is that what might have been intended to remain in the background (how the house looks, someone's son or a naked fiancé) makes its way into the field of vision. The action of Skyping with video, rather than calling on the phone or meeting someone in a more 'neutral' or shared work space, means the background is revealed helping 'explain the conditions of emergence or an arrival of something as the thing that it appears to be in the present' (Ahmed 2006, 38). Through this phenomenological and ethnographic experience Olivia is able to make sense of her fellow board member in a different kind of way. The arrival of her older than expected son in the frame will likely now affect how Olivia understands her fellow board member. Like Olivia, Marae, aged in her late fifties, has also seen the domestic spaces of the people she works with, in this case, two assistants who work for her as part of her small business team. Marae tells me, via a Skype interview:

> There are two assistants now that have rung me on Skype through the iPad and I'm still learning this technology and I could see her and she could see me and I thought, 'Oh gosh.' It felt a little bit funny at first because I have so much communication with them via email and texting regarding work and they've been with me for quite a while and then all of a sudden to be invited into their face and space in their home and I see a husband walking past and kids running around and then I'm wondering about what I look like; it felt a bit strange.

While for both Olivia and Marae the domestic arrangements which they found themselves to be part of felt intriguing and possibly a little strange, sometimes entering the visual frame might be a more serious endeavour with emotional and material consequences as Haylee explains.

Haylee is aged in her mid-thirties and lives with her husband. She says that when her husband Skypes his children from a previous marriage she is not sure whether she should remain in the background or move into view:

> Sometimes I thought 'no, no I wanna be in the picture' because they would often ask 'where's Haylee?' if I wasn't. But I often felt like I was really intruding on their privacy, this is, it's about them, it's about the three of them. It's not about me. So I didn't see anything I shouldn't have but I really, really didn't know whether or not to be in the frame 'cause his computer's static. I use my laptop at home so I can create the background myself, when I'm Skyping I choose, but he can't. His computer is up on a high level and so the whole of the lounge is basically open to the frame, so if I don't want to be in the frame then I actually have to leave that part of the house. I definitely think I felt really awkward quite a few times as to whether or not I should be intruding on that quite personal space.

Haylee continued her story saying that she thought that if she did enter the frame that not only might she be intruding on father-child time and space but also that her partner's ex-wife might be watching the children Skype with their father. Occasionally, Haylee says, she comes into line of sight. When the children Skyped prior to Haylee and her fiancé's wedding she did not go anywhere near the computer because she says that she knows she is a 'trigger' for her fiancé's ex and if something 'had gone wrong she would have stopped the kids from going to the wedding'. Ahmed (2006, 40) poses the question: 'In "having arrived" how does the object [in this case, the image on the screen] become "what," where "what" is open to the "perhaps" of the future.' Haylee 'bringing forth' or arriving on screen heightens awareness of her spatiality, her taking up residence with her fiancé opening up a range of emotions and affects. As Ahmed (2006, 40) notes: 'You bring your past encounters with you when you arrive.'

While this bringing past encounters into the future was fraught for Haylee many other participants reported enjoying this. One particular example is mothers enjoying the company of their adult children by bringing them back (virtually) into the family home. Imogen says:

> If you are Skyping with someone then often they, for example I'm thinking again about children who may have left home, it's kind of putting them back into the family home. They're right there at the table with the rest of the family and life going on around them. Computers aren't just kind of something that tuck away necessarily anymore. They really bring people in.

In previous research (Longhurst 2013) I tell the story of Catherine, aged in her fifties, who likes to stay 'in touch' with her adult son in Australia. In order to do this, the household desktop computer which used to be in a back bedroom has been re-located to the kitchen table (see Royal 2008 on the internet and gendered spaces). Catherine does not particularly like the look of the computer on the table in terms of the aesthetics of the room but she very much likes that her son can now easily appear on the screen, in the dining area and that he, or both of them, can eat or sip a glass of wine and chat for half an hour as though he is actually there – reinstated back in the family home. Catherine describes it as 'wonderful'.

I have returned here to tables. Catherine is not alone in moving her computer from a study or back bedroom to a more central room in the home (see Cassidy 2001 on 'where to place computers on the domestic map'). Given that people now tend to spend considerable periods of time on computers and other mobile devices checking social media, watching television online, sending emails and so on (often synchronously) they tend to want these objects closer at hand – actually in hands, on knees – at home, work and play. Olivia says: 'Our laptop kind of floats around the living areas; lounge, dining room and office.' If people are separated from their objects that connect them to others and to media this distance may register as a threat.

'Nomophobia', a new term added to 'urban dictionaries' over the past few years labels a fear of the loss or impairment of one's mobile phone. Eddie Wren (2012, no page number) in *MailOnline* reports the signs that one may be suffering from nomophobia are 'an inability to ever turn your phone off;

obsessively checking for missed calls, emails and texts; constantly topping up your battery life; and being unable to pop to the bathroom without taking your phone with you.' Many of us today exhibit one or a number of these signs. Kinsley (2013a) refers to 'technicity', that is, how humans and technology exist in a co-constitutive relationship. This is most certainly an example, one cannot exist without the other.

To return though to Fuss's (2004) 'theatres of composition', intellectual labour is animated by bodies, objects and images but other everyday endeavours are also framed in this way. Skype is changing worlds, horizons, the spaces in which we dwell, the directions we face, who faces us. It can disrupt the norms of how bodies and space are supposed to look, smell, sound and feel, that is, how they are supposed to function. Natalie provides an example of this queering from the perspective of a toddler explaining that when she arrived at an airport her grandchild was asked 'Where is Grandma coming from?'. Rather than answer 'New Zealand', the response was 'from the computer'.

A country and a computer were synonymous for the toddler. But even for adults, sometimes when we Skype or use some other kind of synchronous audio-visual platform things are not quite what we are used to, not quite lined up, queered, or as Ewan says: 'kind of mis-matched'.

> On one occasion one of my sons Facetimed. I think we were overseas. Anyway, I remember lying in bed and feeling really foolish, somehow, because I was in my pyjamas in bed and my hair all tussled and he was out somewhere dressed up. It didn't feel right. I felt really vulnerable and a little silly – kind of mis-matched.

This chapter has considered some of the bodily performances involved in Skype with the video, the importance of the visual, what it means to see oneself, to see others and to see different domestic and more public places through the lens. The next chapter (Chapter 5) turns attention towards some of the particular dimensions of what it means for families, friends and loved ones to Skype each other, establishing and maintaining relationships across physical distances.

Note

1 Kiwi is the nickname used internationally for people from Aotearoa New Zealand. 'New Zealanders' also often refer to themselves as Kiwis implying a relaxed but can-do kind of attitude toward life. The name is derived from the flightless kiwi bird.

5 Families, friends and loved ones

Nancy Baym (2010, 1) remarks: 'The digital age is distinguished by rapid transformations in the kinds of technological mediation through which we encounter one another.' As stated previously Skype was first launched in 2003 and initially used primarily for family, friends and loved ones to connect. Over time many people (but not all) have become (at least semi-) accustomed to this new technology. Baym (2010, 1) notes: 'Eventually they [technologies] become so taken for granted they are all but invisible.' Skype has not yet got to the point of being invisible but for family, friends and loved ones using it to connect, there does now appear to at least be some familiarity with how it operates and *feels*. A number of family members, friends and loved ones have to date been using Skype for four or five years, meaning they are currently less self-conscious about the technology and their primary concern is tending toward the relationships, emotions and affects of those involved.

First, in this chapter the focus rests on intergenerational connections between grandparents and grandchildren. Second, the chapter offers an examination of family, friends and loved ones using Skype for special occasions and the ways in which spaces of care are opening up through these occasions. For the most part participants report feeling comfortable conducting these familial relationships. Online and offline spaces coexist comfortably enabling family, friends and loved ones to stay 'in touch' across physical distance. Bodies and spaces – online and offline, cyber and real – are entangled. Throughout the chapter, returning to Ahmed's (2004 and 2006) work on objects, bodies and orientations helps illustrate how the bodies in this project, in relation to family, friends and loved ones, feel able 'to sink' into spaces using Skype.

Across the generations

This section addresses grandparents Skyping grandchildren, namely young children rather than adult children, from the perspective of grandparents. The term 'grandparent' is used here but it is acknowledged that mothers and fathers, and grandmothers and grandfathers often perform different gender roles (see Tarrant (2010) for a spatial perspective that connects old age, men and masculinities; also Tarrant (2013, 2014) and Tarrant and Watts (2014)).

For example, it continues to be mothers, rather than fathers, who typically run households and take responsibility for care-giving. When men retire from paid employment sometimes gender roles and relations can change through developing more intergenerational relationships.

I want, therefore, to broaden the focus on previous research I have conducted beyond women, mothers and grandmothers to also think about grand-fathers. By using the term grand*parents,* I am not meaning to imply that there are not significant differences in gender roles and relations between men and women. Nor am I implying that there is some stable foundation, or essence, to being a grandparent or to the practices surrounding grand-parenting but rather that there are certain dominant discourses such as being loving, affectionate, wise, fun and 'spoiling' grandchildren that have long 'stuck to' grandparenting in a variety of different contexts. Given that there is no essence to being a grandparent it follows that it is a gendered subjectivity that stretches beyond the binary of male/female, men/women and grandmothers/grandfathers.

Also worthy of note, before turning to the empirical data on grandparents and grandchildren, is that it grew out of an earlier project conducted in 2010 (Longhurst 2013). Twenty-four semi-structured individual interviews, two couple interviews and two group interviews were carried out. One group interview involved four mothers aged in their early twenties and the other three mothers aged in their early thirties. In total 35 women took part in the research. All lived in Hamilton, Aotearoa New Zealand. I mention this earlier study because 24 of the 35 mothers involved reported using Skype with video. Although the initial aim of the project was quite broad – to further under-stand some of the social transformations that have occurred in mothering since the early 1990s as more mothers have begun to use the internet – it resulted in an article titled 'Using Skype to mother: bodies, emotions, visuality and screen' (Longhurst 2013). Women using Skype to make video calls, espe-cially to their adult children and/or to their own mothers, usually to facilitate an interaction between grandchildren and grandmothers, emerged as an important theme in the work. I wanted to find out how *seeing* one's child or children as part of the communication affects mothers' feelings towards their children. That work, like this current work, was informed broadly by Ahmed's 'queer phenomenology' but also by the concept of the 'cultural politics of emotions' (Ahmed 2004) rather than 'orientation' (Ahmed 2006). My earlier research concludes that:

> More than half the mothers who were interviewed reported that using Skype with real-time video to see their children reduced feelings of distance. They also reported that 'seeing' their child or children enables them to assess their children's well-being more accurately. In this way the computer screen, as object, by portraying moving visual images, is 'reorientating' mothers' and children's bodies offering a seemingly closer physical and emotional proximity than in the past (Longhurst 2013, 664).

Soon after beginning this project on mothers I became aware that although I had initially imagined the research as being about just mothers it was not possible, nor necessarily advantageous, to bound it in this way. As soon as I started to interview mothers I came to realize that it made little sense to talk just about women's relationship with their children. Mothers also wanted to talk about their relationships with other family members and loved ones, both those living at home and those living away from home. The women who I interviewed tended to slip seamlessly between talking about their relationship with their children, and sometimes grandchildren, and their relationship with their own mothers. By accident, I had stumbled upon a project that involved intergenerationality (Hopkins et al. 2010). In other words, it was not just about mothers but also about grandmothers. The rigid categories that I had initially drawn between mothers and their children, and mothers and their mothers, did not hold (see Figure 5.1).

In recent years there has been growing recognition of the value of an intergenerational approach. Consider, for example, the work of Peter Hopkins

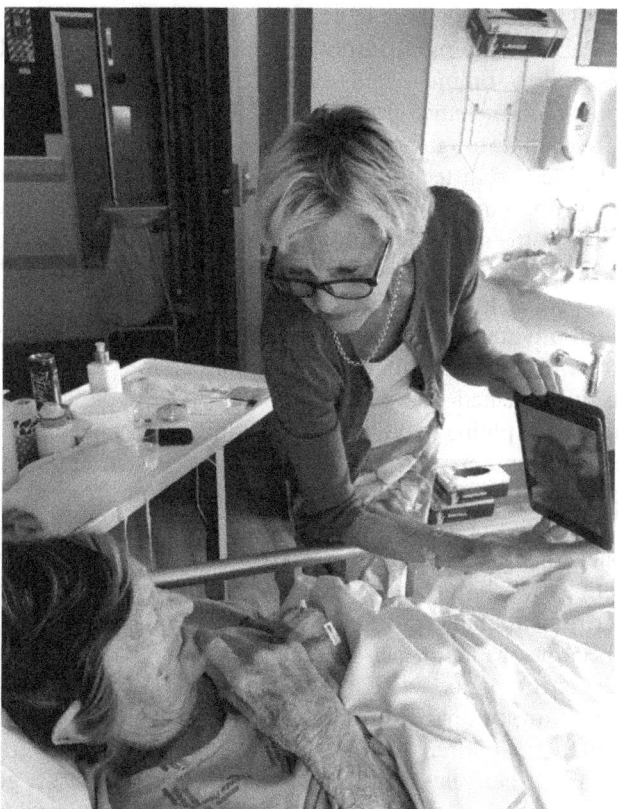

Figure 5.1 Daughter with iPad leans over her mother in hospital as granddaughter blows her a kiss on Skype

and Rachel Pain (2007) on 'geographies of age', Robert Vanderbeck (2007) and Vanderbeck and Nancy Worth (2015) on intergenerational space, Tracy Skelton and Stuart Aitken (Skelton and Aitken 2016) and Skelton and Gill Valentine (Skelton and Valentine 1998) on children's and young people's geographies and Anna Tarrant (2013) on grandfathering. Many of these researchers examine the nature of generational divisions while also recognising the complex ways in which members of different generations are involved in each other's lives (also see Mann, Tarrant and Leeson 2015). For example, Hopkins and Pain (2007) argue it is important to create more *relational* geographies of age. They note: 'Interaction, isolation, divergence, conflict, cooperation and so on all have material effects on the experiences and quality of life of older and younger people in particular settings' (Hopkins and Pain 2007, 289). In the context of Trinidad, Miller and Sinanan (2014, 134) point to the importance of webcam for retaining or maintaining what would otherwise have been 'lost relationships' between grandparents and grandchildren. My research revealed similar results.

In Aotearoa New Zealand and elsewhere internet usage tends to be affected by age in that: 'The younger people are, the more likely they are to use to it, the better their ability, the more important they rate it, the more they create context and socialize online' (Bell et al. 2008, i). But older people are increasingly going online and grandparents and grandchildren are Skyping each other. Children through their digital competency, are able to instruct adults and older people in technology matters, a situation that problematizes the usual 'grown-ups know everything and kids know nothing' scenario (see Correa 2014 on 'bottom up technology transfer within families' and Siibak and Tamme 2013 on 'Who introduced Granny to Facebook?'). Stereo-typically, older people are thought to resist new technologies and yet many are powerfully motivated to learn to use Skype and are embracing it as an opportunity to engage with younger family members.

Participants were asked specifically about their Skype interactions with grandchildren. Fab explains that she Skypes family in France on a daily basis, mainly to keep in contact with grandchildren, but when she first moved to New Zealand in 1992 she had to send a paper letter and that used to take at least a week to 10 days. Fab says that she was completely disconnected from her parents back then and that it 'was pretty much like being on Mars or on another planet'. Now she gets to connect with her grandchildren on a daily basis. This is a radical change.

It quickly became evident when talking with grandparents in this study – eight of the 39 participants – that *seeing* one's grandchildren as part of the communication is important. Not only are many mothers excited by the pro-spect of seeing their children on Skype but so too are many grandparents. Natalie's story about her relationship with her grandchildren is a good case in point. Although Natalie also uses Skype for work purposes and to stay in contact with friends most of the interview was spent talking about the relationship that she has with her grandchildren.

Natalie is a university lecturer in her early sixties. She is an enthusiastic person with a strong sense of the importance of staying connected with family. Natalie, with her husband, migrated from Prague to New Zealand 11 years ago. She has been using Skype for nine years. All her family, including one son, one daughter and five grandchildren, live in Prague. Natalie Skypes them at least once a week. She explained that she has lived most of her life in Prague so Eastern European politics and her family are very important to her. Natalie says: 'The connection, the communication is crucial from many points of view ... I depend on Skype.'

Natalie and her grandchildren have met in person a number of times because she travels from Aotearoa New Zealand to Prague to visit but them but there is, at least for the youngest family member (aged 1–2 years), still confusion as to where Grandma comes from. Natalie explains:

> They see me a lot on the screen when we speak on Skype and on one of the trips when I came I remember ... when I arrived from the airport my son was asking the children 'Where is Grandma coming from?' because they know the name New Zealand. So the older one knew New Zealand but he didn't say, and the little one... I remember that he was holding him so he might have been maybe a year or two years old, he said, 'She came from the computer,' because for him I came suddenly out of the screen.

Toddlers, like Natalie's grandchild, might not be able to relate to a person through a phone call but seeing a person on webcam can prompt recognition.

Natalie continues the conversation describing Skype as 'amazing', saying 'I don't know how people lived without it'. In the past, when there were only telephones and letters as ways of communicating, 'saying goodbye to family knowing that this is what was involved in immigration' must have been very difficult 'but Skype really enables you to be part of the family all the time. We speak for hours.'

Natalie not only *speaks* to family, though, she also childminds via Skype.

> I can tell you my experiences when my granddaughter was really small, she was not one year old. She was able to sit and play with the toys but not to go walking and not crawl. I was babysitting her. My daughter and her partner went to have breakfast in another room, had their own things, and I was just looking at her and if something would be wrong I can shout or call them on the phone if I see she's doing something dangerous or whatever ... I can sit and babysit through Skype.

Natalie continues explaining that Skype enables her to tend to the children in a multitude of ways despite being in another country. She talks about 'having breakfast with her family'. It seems not to matter that they are not sitting at the same table, eating the same food, or experiencing the same time of day/night (9am in the morning in Prague is 9pm in the evening in Aotearoa New Zealand depending on daylight saving). Natalie claims:

> I can do everything. They put the tablet on the table and I have breakfast with them on Saturdays. Today I'll call and I'll see them eating the Saturday breakfast which is bigger, it's when everybody's there coming together, and I'm there on the table with them and we have long conversations and they show me everything. They can bring me the painting that they do or their craftwork, their records from school when they get it, they dance, they sing. I can tell them a story. I have here books now and I can tell them a story through the Skype; everything. It is really enabling me to follow week by week – I can see them growing, I can see them changing.

Sedgwick (2003, 8) in discussing dualisms and 'spatial positionality' explains that for her, the term *beside* is probably the most salient rather than *beneath* or *behind*. This is because 'there is nothing very dualistic about it: a number of elements may lie alongside one another' (p. 8). It is as though Natalie is looking for a way of not being *in front* of her grandchildren on the screen but instead being beside them as they eat, sing and tell stories.

Natalie is clearly an enthusiastic proponent of Skype. When asked if there are things that she is less keen on in relation to the digital communication such as the technology sometimes failing or at times it being intrusive she says:

> I'm always happy to be in New Zealand and be able to be in contact with Prague. And we say so. I tell the grandchildren we need to be very thankful. When something gets disconnected it is frustrating to use, the technology is not perfect, however, you have to appreciate the fact that it is such a distance and you are with them. You are really close … I take the camera. I show them how I put pictures on my wall. I show them all the pictures of the family.

Miller and Sinanan (2014) explain that babies under six months of age are able to differentiate human faces. This includes over webcam. Natalie's grandchildren have been able to recognize and interact with her (smile, vocalize, gaze) at Grandma since they were young but grasping time differences between countries has been more challenging. Natalie explains that her granddaughter:

> can't get it when it is her morning and here it is always night. I have to tell her 'It is midnight, I want to go to bed.' She doesn't believe me so I take the camera and I show her the window and she sees that outside it is dark and she can see the cars going by and then she believes me that it is dark. Yeah, I use it a lot, the camera, to show them.

Transmitting images of the surrounding environment goes both ways. As Natalie mentioned in a previous quote she enjoys seeing things – events, objects, pets – unfolding in her family's household in Prague. She says: 'It was fun last time because they have new dogs … Now, whenever I Skype they show me the dogs as if it is part of the family … all the time I send them presents. So now I send a present also for the dogs.'

As with the project on mothers using Skype to connect with their children, especially those who had left home (Longhurst 2013) the visual aspect of Natalie's contact with her grandchildren is crucial. It reduces feelings of distance between Aotearoa New Zealand and Prague. 'Seeing' her grandchildren enables her to assess their well-being and their growth and maturity more accurately on a weekly basis. In this way the computer screen, by portraying moving visual images of bodies (human and non-human) and objects, acts to 'reorientate' life so that it is seemingly closer, physically and emotionally, than what was possible in the past (see Valentine 2006 on 'globalising intimacy').[1] 'With the kids it's fantastic; you just want to see them and they want to see you, and actually they want to see themselves because in the little picture they see themselves and they look at it all the time,' says Natalie. More than simply seeing the children, however, Natalie also attends to their needs, not their material bodily needs given the distance but mostly their emotional needs. Through Skype she is able to successfully orientate her own body and the bodies of others in her family in ways that enable her to create spaces of care. Natalie tells a poignant story about her granddaughter in Prague having leukaemia.

> And she's just finished her treatment at home, and I used to, all the last year, when she was in hospital I was able to talk to them in the hospital … so I saw the nurses, I saw everything in the hospital in the background … these people didn't know that I was seeing them but I saw them. Actually, it was very helpful. My daughter needs to be with the sick child all the time so, again, I was kind of a babysitter. Yesterday, when I was talking to her, it was morning and my daughter needs to entertain her the whole day; it's really a challenge. So she knows that for an hour her daughter is occupied by me. She shows me how she can dance and she's totally bald. I keep telling her, 'You are beautiful.' At the beginning, even for Skype, she would put on a hat for me and now she doesn't. But being able to speak to her when she was in bed, she could hardly move. She was so weak. She was so in a teary state. When you are here and you imagine the worst but then you see her and you talk to her and we spoke about the clowns that came to entertain her. I saw her. I am just watching. I saw her communicating with the clown. She has lacquer on her hands in different colours and this clown put it on her. So I take notice of these and we speak about basic things to distract her.

Rather than adding to Natalie's stress and sadness about this situation she perceived these interactions with her granddaughter as actually taking a load off her shoulders because she was able to participate in family life (*all of it*, not just the good bits). Natalie not only watched her granddaughter, she also helped her daughter, and in doing so the hospital spaces became less frightening and more everyday (see Sedgwick 1999 on love and illness).

Another way in which Natalie helped the family when her granddaughter was ill, was to travel to Prague to look after her grandson so that her daughter could spend long hours at the hospital. This duty of care also involved Skyping. Natalie explains that although she took her grandson to visit every other day to see his mother and sister, when they were back at home alone he kept missing them and wanted to see his mother.

> So we Skyped to his mother in hospital and he could speak to her, 'Mum, I miss you.' He's three years old. 'I miss you,' and [he'd tell her] a little bit of what he did and then he wanted to go [to the hospital]. He wanted me to take him but it was enough for him [to Skype]. He just needed to see that she was there and she explained to him that she needs to be with his sister because she is very sick and he had to accept it. What can we do? But Skype was very helpful all the time … I am very grateful for Skype.

Natalie being grateful for Skype does not mean that she prefers online to offline communication, commenting:

> Face-to-face is the best …when I look at my grandchildren I want to hug them and they eat breakfast always on Saturday. My daughter prepares fantastic pancakes and I see them eating it and I say to her, 'I want to eat your pancakes.' You want to touch. But it is second best.

I have relayed the conversation with Natalie in detail because in many ways it encapsulates points made by a number of the grandparents namely that Skype is facilitating familial care that extends beyond the spaces of home in ways that were inconceivable only just a few years ago. The research reveals that grandparents are actively seeking innovative ways to communicate with their family members. Skype is an important part of this. Through Skype and other digital media (e.g. Facebook) they are finding ways to offer valuable care and emotional support (Tarrant 2014). Grandparents can be both present and absent in the lives of their grandchildren. They come in and out of their daily routines on screens and in real life. Children may react differently in different situations as their perspectives in time and space changes.

To return to the work of Ahmed, she says: 'Comfort is the effect of bodies being able to "sink" into spaces that have already taken their shape' (Ahmed 2004, 152). All of the eight grandparents who were interviewed are comfortable sinking into the spaces created by Skype to connect with their grandchildren. It has been really important for Natalie to actually *see* her grandchildren. She experiences them as immediately present in the space of her home in Aotearoa New Zealand and their home (and a hospital) in Prague. The sound of their voices but also the look of their bodies, pictures, toys, pancakes and pets are important to her. Grandparents like to *see* their grandchildren. As I have pointed out elsewhere (Longhurst 2013) Henri Lefebvre (1991) notes, visuality – the eye – is often prioritized over other senses such as taste, smell and touch, and even hearing, in 'modern', Western, capitalist societies. It is thought to provide something of an epistemological guarantee (Jay 1993).

Barbara and Henare, a couple in their mid-fifties who were interviewed together, are also keen to stay in touch with a grandchild in Australia. Barbara says that when they Skyped him:

> He bonded with me almost immediately and was really awesome and fun and interactive but when we got over there [Australia] 'cause we'd only just met him and he was a year old he looked at Granddad [Henare] and he was like 'Who the hell are you? You're a big fella. I'm not coming near you'. And he didn't go near Granddad for 20 minutes/half an hour and just played with me. And then all of a sudden he did, he approached you, didn't he? [Barbara says to her husband beside her]

Granddad, in the flesh, felt bigger to his grandson than he had on the screen. On the screen the child felt comfortable with Granddad but when Granddad arrived 'on the ground' in Australia things changed. In a similar situation, Bryan is retired and lives in Australia. One of his granddaughters lives in Aotearoa New Zealand. He tells me that she really enjoys Skype sessions that involve 'show and tell' (a child displays an object then talks about it).

> 'Show and tell' is very good for Skype. I think getting back to voice calls they tend to be often well meaning grandparents asking questions like how are you, how is school and what did you do today? In a way those are the last questions that kids want to be asked; they're tired of those questions.

Bryan says, referring to his granddaughter showing him a recent painting she'd done, 'You can see things in a more dramatic way and more personal way.' Younger and older children often present to grandparents drawings, homework, pets, bedrooms, gardens, baking, new school uniforms and so on. They do not tend to focus on the technology per se but on what they are saying, doing, touching, showing, and most importantly, seeing (see Figure 5.2). Miller and Sinanan (2014, 121) argue that there is 'an important symmetrical relationship between the oldest and youngest members' of families. This is partly because it is experienced with a degree of immediacy with another person rather than as a technology.

There was also a lot of 'show and tell' for Queenie. She describes her use of Skype over the past two years as 'intensive' on account of her wanting to keep in contact with family. Queenie goes to Australia to visit her grandchildren every couple of months. She says it is 'very normal' for them to Skype.

Figure 5.2 Nephew shows his Great Aunt his half-eaten apple while she is a passenger in car

They come and talk to me and they interact with me just like you and I are sitting here talking [the interview was conducted face-to-face]. They show me their nail polish and their hair and they say, 'Look at this Nana' and they'll go and bring something across or they'll do a drawing and bring it up to the screen. So it's an interaction thing, it's not just a matter of sitting and talking, it's that they're interacting and as if I was there, they're doing exactly the same things. If they've got a new scooter or a bike they say, 'Watch me' and, 'Watch me jump' and, 'Watch me on this.' If I was sitting there they'd be doing exactly that sort of thing.

In the next chapter on using Skype at work the words 'weird' and 'strange' appear with some regularity. In relation to family, friends and loved ones, however, the words 'normal' and 'natural' were more common when participants talked about their grandchildren. Imogen notes there 'are a lot of kids now with grandparents in other places'. She says her stepdaughter has got 'a little boy who is nearly four and he's met me twice but also he's seen me on Skype so its normal for him to have this relationship with a grandparent, a person who is not always there, or not very often there, but can be like there talking to him while he's eating his breakfast'. The words 'normal' and 'natural' emerged when bodies are orientated towards the objects of everyday life, kitchen tables, plates, bikes, drawings and it feels comfortable.

Eva, a homemaker in her late fifties, provides a useful example of how seeing her grandchildren via Skype feels 'normal'. She does not mind if when she Skypes other people her webcam does not work but with the grand-children she really likes to able to see them and for them to see her. She says that it is much better than just talking on the phone and if it was not for the visual she would not bother. Eva says:

> I will give you an example. Two weeks ago I got back from spending nine days with my grandsons. Well the very next day I Skyped them because it was like I was seeing them the next day and yesterday I Skyped them and I had brought some dragon fruit and dragon fruit are the most beautiful looking fruit so I showed them the dragon fruit on the Skype camera. Then I cut it open and I showed it to them cutting it open and the day before that they were not well and so I Skyped them and [name of first grandson] said to me, '[name of second grandson] not well today Granny, I not go to the park today' and so I was able to empathise with that and ask him how he was feeling … [name of third grandson] was throwing a wobbly [Australasian slang for tantrum] so he was grizzly most of the time but then I said to my daughter, 'Why don't you put some oil on his spine or put some oil on his feet?' So she just had a couple of empty bottles of oil and she gave him these bottles of oil and he goes, 'oiler oiler' and here he was playing with these and he cheered up. I don't know what made him cheer up; and then by the time we were nearly ready to go he started waving and he started looking and he said, 'Granny' and you know when I got on that Skype call they were both a bit grumpy and [name of third grandson] was still grumpy for like the first five minutes and then by the time we got off he cheered up and they were both quite cheerful and they remained so for the rest of the afternoon according to my daughter. So in that way I don't know what would have happened had we not Skyped … You don't know how you are going to find your grandchildren. You don't know when you Skype them whether they are having a busy day or a grumpy day or they have been out or they have banged their heads you know, or they are watching Play School [edu-cational television show for children]. You just don't know what they are doing so you can Skype and sometimes they don't pick up and sometimes they do. Or they could be eating a meal or they could be helping cook; all this sort of stuff is going on and you get to engage with them through any of it.

Eva says that because of the visibility offered through Skype, her grandchildren are 'closely connected' to her.

Again, to filter this story through the lens provided by Ahmed (2006), what is relegated to the background with a phone call – a child scowling or sitting quietly watching television – comes into view with Skype, enabling those involved to feel more orientated even if the emotions being displayed are not happy emotions. The affect may be anger or sadness but they are nevertheless orientating for the bodies concerned. The phenomenology of Skype when children are involved, *not* sitting still and focused in front of the webcam but moving, playing, fighting, displaying objects and kissing screens means that many things, including multiple backgrounds, are often revealed. Attention may shift, 'wander' and 'turn toward' (see Ahmed 2006, 29) the unexpected. Children sometimes direct their attention to objects, such as hitting or hugging a screen, in unexpected ways, and yet clearly for the most part there is a mutual enjoyment in this. Eva says her grandchildren put their heads right up to the screen and she can see their hair, and then they go 'kiss, kiss, kiss, kiss, kiss – love you granny'. She really enjoys this.

Despite the many positive comments that grandparents offered in the interviews, occasionally there was a glimpse of how grandparents, parents and children have had to learn how to use Skype together over a lengthy period of time. It has now been used for the purposes of family connecting for more than a decade and has come to feel normal, natural or ordinary for most, although this has not always been the case. Travis, who is in his late sixties and lives with his wife, enjoys Skyping their granddaughter. He remarks:

> I have used it quite a bit and I just find it very good. When we first started using it it was the new frontier and it's now, it's sort of starting to get a wee [little] bit just normal. I think that's the thing with it; when it first started it was just great and wonderful but I guess it's like everything once you get a wee bit used to it, it's just back to normal.

For Travis, normal appears to feel a little boring. For most participants, though, normal felt comfortable, something one no longer had to think consciously about. Children, parents and grandchildren have all had to learn how to connect, that is, how to orientate themselves around this new communication technology and learning how to communicate can be stressful.

Debbie explained that her parents initially felt quite anxious about not getting to see much of their granddaughter because she lives in a different country and often she did not want to talk with them on the phone. Debbie sensed that they were quite hurt by this and kept saying to them: 'But she's only a little kid, she doesn't really like telephones.' It was at that point that they began to use Skype and it solved the problem but Debbie explains that she had to instruct her daughter in order to make it work with her parents.

She's [Debbie's daughter] quite patient with my parents and I can see that for them, it's quite important. As she's grown older and more mature she's quite sensitive to that so now she sits still for quite a long time and listens as well. We've [Debbie and her partner] encouraged her to ask them [Debbie's parents] questions so it's not just all about her performing which I don't like, grandparents wanting grandchildren to perform. I say to her, 'Why don't you ask them what they've been up to?' Then they have a real conversation. I think, visually, they can show her things; they show her a letter that she might have sent them, they can show her that it's been received at the other end. There's an interesting kind of, almost like a little wall that people can pass through in that sense, when it works. When it doesn't work, I don't like it at all. When it does, it's nice.

Debbie's comment about Skype being 'a little wall that people can pass through' conjures up images of a tear in the time-space continuum making it possible for grandparents, parents and children to teleport, or transfer their bodies from one space to another without actually traversing physical space. The worlds of the children, parents and grandparents fold.

It can take time, though, to get a feel for this 'passing through the little wall' for some family members, especially the very young. Fab says she puts her iPad in front of her baby when she is having breakfast so she can see her grandparents on the screen. The baby has become accustomed to this and when she hears the Skype ringtone she now looks at the iPad and smiles. Quite often she says 'hello' followed by 'baby, baby' because her grandparents look small on the screen. Fab, her mother tells her, 'no, not baby, you baby'.

As a counter to the many positive views expressed by the participants about Skype, Ariana, aged in her early twenties, says that her 'Nan' will 'not go anywhere near Skype' because 'it freaks her out'. She continues:

I think it's only my generation that appreciates it [Skype]. My Mum and that, they'll do it if they have to but they don't really want to. They're more about ringing you up on the phone or something. They're not really going to jump on Skype. If I say, 'Go on Skype' she doesn't know what I'm talking about or she thinks it's too hard. You've got to be well-versed to know that it's easy otherwise you'll never really use it. People just have all these pre-conceptions that it's hard.

Despite Arianna's comments the majority of participants illustrate through their narratives that grandparents as well as other family members are developing and maintaining emotional and familial links across generations using Skype. As was the case with mothers and children in my earlier round of fieldwork (Longhurst 2014), grandparents feel it is vitally important to actually *see* their grandchild or grandchildren as part of the communication. This makes a powerful difference in enabling their bodies to 'sink' (Ahmed 2004) and sync into home spaces that feel comfortable. Of course feeling comfortable does not necessarily occur from the outset. Debbie explains that she thinks that her daughter and parents being able to Skype is 'fantastic', however, she then adds: 'But, I have to say, and I now think I'm a lot more used to it, it's taken me a while to feel as comfortable.' Comfort was not immediate. When they first started Skyping approximately three or four years ago Debbie remembers saying: 'Dad, there's a button here, turn on your speakers.' She says that there was frustration at trying to make it work. She also recalls that her daughter, who is now seven, was just a pre-schooler and 'more restless'. Debbie notes: 'so it was really hard for her to understand that they wanted to see her, they wanted her to talk to them, they wanted her to look at them and interact with them through Skype'. Now that her daughter is older and they have all grown used to Skype it is 'a bit different'. Debbie concludes: 'There's been a change over time in terms of the family Skyping.' The sense of frustration that Debbie used to feel as she tried to ensure that everything was working well – the sound and image, and that the interactions between family members ran smoothly – have dissipated over time as everyone has grown more used to the medium.

Tarrant (2014), in her research with 31 grandfathers about communication technologies, also revealed that they actively sought to maintain close relationships with their grandchildren through the use of technology. Due to increased mobility families are more likely than ever before to be geographically spread. Skype then is playing a major role in connecting families. Also, as people become older they often face challenges such as loneliness (Yeh and Sing 2004) and impaired mobility finding it more difficult to keep up with their social contacts. Being able to reorientate their bodies, therefore, around a computer or device that enables them to connect without having to physically get somewhere can be enriching. Families, friends and loved ones being able to connect across distance for special occasions is proving to be enriching for many.

Special occasions

In 2015 the popular American television mockumentary *Modern Family* screened an episode titled 'American Skyper'. It is graduation day for daughter Alex (Ariel Winter) and the whole family has gathered to celebrate. Unfortunately, however, Phil (Ty Burrell) is stuck in Seattle and so has to be the proud father from afar. Therefore, Phil decides to take part via Skype (an iPad mounted on to a mobile robot). Skype works well for the most part, enabling Phil to be part of the unfolding of the family dramas, but in the final instance the technology fails and Phil can hear others but no one can hear him prompting a number of humorous moments.

I describe the 'American Skyper' episode of *Modern Family* because popular culture or 'vulgar geographies' (Kinsley 2016) are currently rife with accounts of using Skype for special occasions. Searching 'Skype Special Occasions' in Reddit generates 32,500 results. It is a popular topic of discussion. One of the special occasions often discussed is weddings. The following post seeks advice about 'Sick grandparents who can't travel' to a wedding. The bride-to-be writes: 'Both me and my FH [future husband] have grandparents who can't travel ... We are both close to our grandparents and want them to be a part of our life changing event but I'm not really sure what to do because there is no way to have them both attend.' Having family members live in different locations – Wyoming where one set of ailing grandparents live, Pennsylvania where another set of ailing grandparents live, and Arizona where the couple-to-be-wed live – means that the decision of where to have the wedding ceremony is complicated. The bride-to-be seeks advice, asking other Reddit users whether they need to have two small private ceremonies for each set of grandparents, or whether it would be 'impolite' to in fact have just one ceremony some-where entirely different – Colorado where the bride-to-be's parents live. She says, we could include our grandparents in this ceremony using Skype.

A Reddit user who describes herself as a 'Wedding Planner', responds to the post explaining: 'I work with mostly Indo-Pakistani brides, and most family is overseas and too old or sick to travel. Lots of family who cannot attend will skype in and have a designated skype-host, someone who is actually at the event itself and can walk around. It means the world.' Referring to tears of joy, the wedding planner says this offers an important way of sharing despite physical distance. She continues that many brides (including herself) eventually return to 'the motherland' and when this happens the couple or their families tend to host a dinner to celebrate even if the wedding was quite some time ago but in the interim Skype offers a useful solution enabling families to be (virtually) together.

While Skype is being used as part of everyday familial and intergenerational interactions it is also being used for special occasions such as graduations, weddings, birthdays, anniversaries and Christmas. Entering 'Skype weddings' into the Google search engine returns more than a million results. Topics include live-streaming weddings, the legality of Skype marriages, Skype marriages for immigration and trafficking purposes, saving money with a Skype wedding, Islamic weddings (nikah) via Skype, and consummating the marriage.

This is perhaps not surprising given that there is often a lot to 'manage' in trying to (re)orientate bodies and objects materially and virtually, to 'line them up' (Ahmed 2006), in order for Skype to work effectively for a special occasion. In one of the interviews a couple, James and Margo, who had recently migrated from the United Kingdom to Aotearoa New Zealand told a story about getting out of bed at 2 am, dressing up in formal attire, opening a bottle of champagne, and turning on the computer in order to attend via audio-visual link (in this instance Google Hangout rather than Skype) their friends' wedding. The wedding was taking place outside in Majorca, Spain. Unfortunately, the link failed and James and Margo were unable to see or hear the bride and groom. They did, however, stay up until 3 or 4 am (Aotearoa New Zealand time) chatting online and having a drink (virtually) with some of the other guests located in Whitby, United Kingdom who were also online and unable to see the bride and groom due to a poor Wi-Fi connection.

Beverly, another participant aged in her late twenties who has been using Skype for six years, says that she just 'can't imagine Skyping for a wedding. If it was an immediate family member yes, but I have close friends who have got married but I guess it's like the interrupting thing, I would sort of assume that they were busy on that day ... and didn't want to hear from me!' Beverly comments that she would prefer to send a message, email, or 'even a letter' so they could open it at their convenience. These two opposing viewpoints illustrate that not everyone is in favour of attending weddings via Skype.

This section examines some of the emotions and affects of using Skype for special occasions. It addresses in what ways Skyping for a special occasion is similar to or different from more frequent Skyping. On special occasions webcams are sometimes left on for many hours (although this can also be part of people's daily Skype routine with family members). Sometimes the webcam might be stationary, other times people might use a mobile device such as an iPad, tablet or phone. The aim is for friends or family members who are often in another state, region or overseas to feel included in an event. Christmas and birthdays seemed to be the most Skype celebrated events for participants in this study.

Beverly says that she always Skypes her immediate family members at Christmas, commenting: 'That's really important for us. I think it would cause major family problems particularly at Christmas if we didn't hold up our end of that kind of implicit deal.' When I asked Kieran, a university student in his late teens, if he had used Skype to be with others on a special occasion he singled out Christmas saying: 'Yeah, to see other members of the family and open presents and those sorts of things. Or Russian New Year, sort of Christmas stuff, Skyping the family that's over in Russia.' Opening presents while Skyping enables family and friends to see reactions to gifts that would not otherwise by possible. A few of the participants, beside Kieran, also mentioned opening presents. James (who was interviewed with Margo) recalls:

> Neil and Ellen opened all their presents whilst we were Skyping and that was really, really nice because when someone's doing that you're just observing anyway a lot of the time and just laughing at how excited the little girl was getting and tearing the paper off frantically which just translates well. I thought that was really nice.

James and Margo also Skyped their friends in their home country the United Kingdom on New Year's Eve. Given the time difference this posed some challenges. They explain that they celebrated New Year's Eve in Aotearoa New Zealand first and then at midday their friends got in touch with them just 10 minutes before midnight in the United Kingdom. They enjoyed this very much because there was a large gathering of people, including children. James and Margo were involved in the countdown to midnight (but midday for them in Aotearoa New Zealand). Margo says: 'I'm getting a lump in my throat now thinking about it. It was lovely.'

Travis, when asked if he Skypes with people for special occasions, says

TRAVIS: Birthdays; quite often for birthdays and anniversaries and things like that ... Mainly with family. We don't use it a lot with friends; most of our friends are in Christchurch of course so we just pick up the phone or if they're out of Christchurch you usually just talk on the phone. Those calls aren't as sort of intimate as family calls and I think this sort of media works better with family. I guess if you're used to using it for business or something like that then that's good but I find it best for families; I mean it doesn't matter then if you're still in your dressing gown or something like that, it doesn't matter does it. But on the phone it doesn't matter what you're wearing.

INTERVIEWER: So when you've used it for special occasions like birthdays, what is that like?

TRAVIS: It's good. If it's [child's name] birthday then usually we'll get on and sing happy birthday and that sort of thing. She can show us what she's got for her birthday for presents and things like that and also at Christmas time and things like that; if we can't be together then this a very good medium for talking to family in those special times.

Beverly too says that it is important for her to connect with family to celebrate special occasions:

I feel unless there's a really good reason that we're doing something else on someone's birthday, it usually feels a little bit strange not to Skype with them. I mean, I try and remember to Skype my parents for Mother's and Father's Day but the dates of that are all different in different countries – I get totally confused about when that is and they don't often know when Mother's Day is in the country that I'm in anyway ... So that doesn't always work out but we try and do that.

Brittney says that Skyping on special occasions 'only works well if my Dad is there; Mum doesn't know how to'. She likes, at events such as birthdays, to see lots of people on the screen at once. Sometimes, however, although Brittney enjoys these occasions it can prompt her to miss family more by highlighting some of the physical things [such as sharing a meal] that she feels she is missing out on.

We use Skype with our family in Thailand for birthdays, for [grand-mother's name] who is our Grandma and that's always fun when you're doing it and then when you get off you're like, 'Oh, I don't get to eat the yum food,' and don't get to be with them obviously; they're all getting ready to go out to her favourite restaurant and some of us are just stuck in New Zealand.

Fab makes a similar point explaining that last weekend was her son's ninth birthday and so they Skyped her parents (his grandparents) who were involved in blowing out the candles with him. She notes 'these are occasions that you don't want to miss' but it 'can highlight the fact that you're not together because you can't hug each other or you can't eat a piece of the birthday cake or whatever, but I think at the same time it's quite good because otherwise you wouldn't be able to be in the occasion at all. So there's two ways to look at it.'

Debbie also feels sometimes after Skyping family that she is missing out (she describes this as 'distancing'). This was not just on special occasions. Debbie says sometimes, after turning off Skype, she has a feeling that everyone will just go back to their own lives and that her Mum and Dad will see her sister and her family that day and that they will all go out for lunch. This leaves her feeling 'really annoyed'. Debbie says she has 'this sense of them having this other world which I would quite like to be able to jump into.'

Mother and daughter Keri and Francine, who were interviewed together on Skype, say that they Skype for special occasions.

KERI: Yeah and then family occasions like Christmas or birthdays everyone gets in on the action.
FRANCINE: It gets passed around so you can see everybody.
KERI: And that helps people who don't have access to Skype like grandmothers to utilise it and see the grandchildren.
FRANCINE: I think they appreciate that as well.

Whereas Keri and Francine were able to remember quite clearly Skyping on special occasions, James and Margo had difficulty trying to try recall together when they had used Skype for different celebrations. As recent migrants Skype has become such a regular part of their everyday lives for connecting with family, friends and loved ones back in the United Kingdom that the Skype sessions sometimes blur for them, just like events in real life blur.

MARGO: We had one [a communication via Skype] at Neil's birthday party didn't we? That was an ad hoc Skype and it was a Sunday morning for us and a Saturday night for them and it was on a mobile phone which is never as good as what it is on a laptop.
JAMES: Yeah, it's a big difference.
MARGO: So the phone was just going around and we were saying 'hello' to everyone, which was lovely, but I don't think they could hear what we were saying and there wasn't really a conversation, it was just a lot of hooray!
MARGO: And Christmas Day?
JAMES: New Year's Eve? Christmas Day, you did present opening?
MARGO: No.

JAMES: Did we not?

MARGO: Did we? Was it Neil and Kate's? It might have been Pete's. We might have opened someone's present.

JAMES: No he opened his, we did open a present with someone via Skype didn't we?

MARGO: Our Christmas morning, it was still Christmas Eve for them.

Margo and James are very comfortable with the medium of Skype, telling me that because of the 12 hour time difference they often Skype from bed (it is night time in New Zealand and morning in the United Kingdom). James says that his Mum 'likes to do her hair and all that business' but he and Margo do not bother.

As Skype increases in use and familiarity with family, friends and loved ones, to the point that it is beginning to feel more 'natural', people are beginning to engage with it in ways previously thought impossible, not only by keeping it on – livestreaming – for longer periods but also using it for a wider range of occasions. One such occasion is funerals. While only one of the 39 participants in this study had attended a funeral using Skype, many around the world have done so.

Searching 'Skype funerals' in Google returns approximately 150,000 hits. 'Online funerals' returns approximately 11 million hits. Noticeable in the list of results are undertakers advertising that they can arrange Skype both so that family members can participate in the funeral arrangements and/or actually attend the service online. The *Mail Online* headline 'Is it on pay-per-pew? Funeral to be live-streamed online as undertakers move their services into the digital age' (Smith 2014, no page number) reports on this issue, explaining that funeral homes are now streaming services online so that people who are unable to attend, for example, ex-pats, soldiers and long-lost friends, can observe and say their good-byes from a distance. In Wales, since 2012, more than half the crematoria have been fitted with webcams. There are even higher proportions in England. 'In some cases, funeral directors will make footage of the service available for 30 days by uploading it to their website ... While some families request professional filming of services, others have asked directors to set up Skype while the ceremony is carried out, experts said' (Smith 2014, no page number). Skype is enabling people to farewell their friends, family members and loved ones in new ways. How people think about funerals and grief is changing. New elements are being incorporated in long-standing traditions. There is often uncertainty about how best to do this.

It is perhaps not surprising therefore that there are numerous fora, online posts, websites, blogs and threads on Reddit where people exchange information, for example, on how to live-stream a funeral, share experiences of having attended a funeral using Skype, discussions about whether it is socially accep-table to do this, and much more. There is a growing industry in countries such as the United States, Canada, United Kingdom, Australia and Aotearoa New Zealand, where mortuaries are setting up internet services to enable families to take part in funerals from a distance (Smith 2014).

In a blog titled 'Figuring out how to mourn in the age of Skype' (Stoker Bruenig 2014), a student who completed her master's dissertation in the United Kingdom before returning home to the United States, explains the grief she felt when her supervisor less than a month after she went home was killed in a car accident. The student's grief was real and yet she had no one to share it with. None of her family in the United States knew her supervisor. Elizabeth Stoker Bruenig (2014) points out that as more people live outside their country of origin, travel globally and study abroad it is increasingly likely that some will have to mourn alone at various stages in their lives. We need to figure out, therefore, how to mourn from a distance. The student says: 'It's hard to be a member of the first generation of online mourners, because there's no book to tell you how you should act – something that's usually a huge comfort in times of loss' (Stoker Bruenig 2014, no page number).

Brianna, aged in her early twenties, was the only participant in this research who had taken part in a funeral (tangi)[2] using Skype. Brianna tells her story

BRIANNA: After that [a large celebratory Christmas event that included family in Aotearoa New Zealand and Australia] we knew it worked so we used Skype for every single event we'd have like birthdays, family lunches, even funerals – tangi. Very, very contentious to use something like Skype [for tangi] but you don't film the body. There's protocol around it. It was just to film the gathering. We've used it in that sense as well. It was very problematic because you had Skype and then you had a conflict with custom. Everything that we had for a hundred years, so technology is hitting a lot of those boundaries we've put up.

INTERVIEWER: That's really interesting. How did you negotiate that?

BRIANNA: We negotiated it in the sense that we talked through what we were going to film. It was like going for ethics all over again but it was harder because it's your own people. It's your own elders. We're big believers of karma and things. If you don't do things right or correctly then it will come back to you and it won't be so good. We had to go through the elders of the marae [meeting house] in exactly what we were going to film and when. Also stuff like can we talk on the marae? Can we film in the wharekai [eating house] where we eat? Those kinds of things. It's very shaky ground I guess 'cause it's hard to detail what you're going to film and what you're going to see.

INTERVIEWER: It was just a one-off constant stream? It wasn't actually being recorded was it?

BRIANNA: No, it was on Skype, yeah. You choose who you want to see it. I think you could upload your Skype name to the people you want to see. I'll try and organize it for my Nana when she dies because she's quite an influential person in the Māori world. If she were to pass away, because she's quite sick right now, we'd need to find a way to stream it 'cause there's a lot of family overseas. It does conflict with a lot of traditional beliefs.

INTERVIEWER: How many times have tangi been filmed now or Skyped?

BRIANNA: Only about twice.

INTERVIEWER: Okay so still relatively new. How long ago was it that you Skyped the very first one?

BRIANNA: The first one was done when my cousin passed away, two years ago. She was 14 and she actually died in Perth. It was from an aneurism so ended up flying her back home and there was her friends and her family were in Perth because they couldn't all afford to come back. If there are circumstances where you have to then they'll let us but if you don't have to they won't let us.

INTERVIEWER: It's very, very interesting. Do you know of anyone else who is doing it?

BRIANNA: Yeah. It's definitely been pushing the envelope. It's verging with the future and the way that we thought it was; it's changing. There's this big message out there that you gotta get with technology before it runs you over. We hear that a lot. You either get with it and embrace it the way you want it and make your own ground rules around it or somebody's going to come in and do it where it's going to be wrong or considered wrong. We either need to get with it and make our own rules and guidelines around it or we're going to get lost ... There are some forward things I am pushing because I'm a trustee back home for all our iwi [tribe], hapū [number of extended family groups], marae issues. I've brought the technology with me and given tutorials on how you use Skype and Facebook, to the elders, and saying that there's important ways that we need to be using this technology. It just changed the way we do things, but it will but there's ways we can do it that's conducive to us and that it can grow us. ... I'm of the view that it brings people together because we would feel further apart without it. Without Skype, we wouldn't see each other ever. Unless they came home or we went over there, then we would never see each other. It's not really possible to go over to Australia. We all have jobs and things like that. It makes the world not feel so big.

What Brianna's story makes clear is that Skype is changing how people do things including grieving, death and burial. Her story also makes it clear that it matters exactly which people – Māori, Pākehā (white European), Catholic, Muslim, non-religious – we are considering in relation to using Skype for a funeral or grieving. How Māori engage with Skype for tangi may well be different from how Pākehā engage with it for funerals, for example, not including the body in the frame.

'Sinking' into the spaces of Skype

Some people have now been using Skype to connect with family and friends for more than a decade. When technologies are new it is often difficult to be able to 'sink' into spaces with them. Miller and Sinanan (2014, 3) start their book *Webcam* with a vignette from a participant who says: 'You know what men are like, they are so impatient with technology. They get so easily frustrated and angry. And to be honest, webcam in those days was pretty crap, kept cutting off, out of focus, just starting a conversation and it goes wrong.' Skype's user-friendliness over the past 10 years, though, has improved, to the point where many grandparents and grandchildren now seem very comfortable with it. It has enabled grandparents and grandchildren to establish relationships that might otherwise have not been possible. All of the grandparents who were interviewed for this project enjoyed using Skype to communicate with their children and grandchildren but many parents also reported enjoying facilitating and seeing their children and parents interact online. They liked it that through Skype, whether it be daily, weekly or more infrequently, grandparents and grandchildren could establish a relationship despite physical distance.

Grandparents often appreciated that children were not just sitting still facing the camera but that they were having breakfast, playing with toys, showing them objects, or even 'throwing a wobbly' (as mentioned by Natalie). Both the children and adults for the most part appear to be un-phased by the technology as emotions and affects transmitted through wires from space to space with little effort. For these grandparents calls with video are a form of interaction that is comfortable, like 'sinking' into a space that feels at home. The hardware and software work together to stretch horizons, taking children and adults into each other's domestic spaces. The days of sending letters, tape cassettes with recorded voice messages, postcards and gifts used by families in the past as ways of connecting seem distant and not always fondly remembered by participants. Grandparents and grandchildren are able to love each other, creating an affective bond through a technical interface. Sedgwick (1999), alongside her work on queer theory, also writes on love, arguing for the possibilities of pleasure, love and hope which she says are never autonomous but always relational, complex and multilayered.

In addition to the love shared between grandparents and grandchildren, families, friends and loved ones have increasingly over the past decade been using Skype for special occasions, not so much for weddings and funerals (yet) but certainly for birthdays and annual events such as Christmas. Sometimes there are technical problems in making this work successfully and other times not but the protocols surrounding formal events are uncertain when people are attending virtually. Nevertheless, people appear to be becoming increasingly comfortable, or orientated towards using Skype for these occasions.

Ahmed (2006, 51) writes: 'To orientate oneself can mean to adjust one's position, or another's position, such that we are "facing" the right direction: we know where we are through how we position ourselves in relation to others.' People are working out how to position themselves for these Skype calls with videos, when to dress up, when to stay in pyjamas, which objects to allow into the frame, which ones to exclude, when to sit still in front of the camera and when to move. Skype was originally set up as a platform for connecting with family, friends and loved ones when distance separated us so this is what we have been using it for the longest. When we get it wrong with family, friends and loved ones the emotional stakes can be high but most participants tended to feel that it was okay, 'just family' and so not as bad as when things with Skype go wrong at work or for a job interview (as discussed in the next chapter).

Ahmed (2006, 51) reminds us that objects can 'take the shape of the bodies for whom they are "intended"' and that they are 'made for some kinds of bodies more than others'. Not everyone has access to (Janelle and Hodge 2000) or finds computers, devices, cables, keyboards, plugs, screens, software and internet connections easy objects or tools to use but many people are becoming increasingly familiar with these things and the objects/tools themselves are continually being altered in an attempt to make them increasingly take the shape of the bodies for whom they are intended. Examples include simplifying a computer interface or changing a computer mouse so it can be used by someone with arthritis or improving screen visibility. This means that in using Skype, at least for the purposes of connecting with family, friends and loved ones, most of the participants in this study, most of the time, are feeling 'at home'.

This does not mean that there have not been queer things at play. To remind readers, Ahmed (2006) argues that to make things queer is to disturb the order of things. Skype has disturbed the order of things by way of technology, objects, and bodies seeming unfamiliar. We have had to get used to seeing ourselves in the small box off to the side, we [including the elderly] have had to learn new technological skills, babies and toddlers have had to work out how to differentiate between representations and reality, parents have attempted to manage the bodily performances of children, unexpected objects and people appearing in the background – but we are orientating ourselves towards this new queerer way of being. For now, for the most part, friends, family and loved ones report feeling comfortable using Skype. Using the technology for work or employment purposes, however, is a different story.

Notes

1 While I am stressing here many of the positive dimensions of children and older people interacting online, for more information on some of the less positive dimension of children online, and parents' concerns about their use of the internet see Valentine and Holloway (2001b and 2001c). This offers a different perspective on children's use of the internet for purposes other than connecting with their grandparents.

2 Tangihanga, or more commonly, tangi, is a traditional Māori funeral rite held on a marae (area outside the meeting house). It is still widely practiced in Aotearoa New Zealand and provides an opportunity for communities, iwi (tribe) and whanau (family) to gather and express grief for the person who has died. Formal rituals differ from iwi to iwi in how they honour those who pass but all are treated with reverence. Tangihanga often take three days with burial on the third day.

6   Skype for work
'A bit weird'

The subtitle for this chapter, 'It's a bit weird', is a quote from one of the research participants, Natalie, who is describing being interviewed for a job using Skype. In fact the adjectives 'weird' and 'strange' surfaced a number of times in participants' narratives about using Skype at work. Searching the transcripts reveals that the 39 participants used the word 'weird' more than 50 times, not always but often in relation to work. While it seems to be becoming increasingly 'normal' to Skype friends, family and loved ones, Skyping colleagues, or doing a job interview using Skype still seems for many to be unusual, disorientating, or to use Natalie's and several other participants' word, 'weird'.

This chapter addresses a series of questions including: does seeing a colleague(s), situated in a particular space and time (such as in their home instead of in their office), rather than phoning or emailing make a difference in some way to the interaction, the orientation, and if so how? How do workers who are based in the field, for example, on a reasonably remote island like one of the participants feel about Skype? What about employees who are expected to stay logged into Skype all the time just in case their employer wants to call them? Employers' and employees' visual presence at a distance may not necessarily be benign or positive. In short, this chapter addresses what kinds of impacts Skype might be having on employers, but mainly on employees, in a range of workplaces and whether workplaces are acquiring a new shape given the increasing number and intensity of interactions happening via a screen.

Until recently Skype with video has been used to connect just two parties. Lync was the platform commonly used to connect multiple parties. In 2015, however, Lync became Skype for Business. It is likely that in the future more business and organizations will use this platform but a number of participants in this study discuss other audio-visual platforms given that Skype for Business is still quite new. For the purposes of thinking through bodies, spaces and screens, however, it is in some ways irrelevant. The effects of people attempting to inhabit new audio-visual spaces, whether it be Skype or another platform, are much the same.

To begin this chapter I examine some of the narratives of the 39 participants about being interviewed for a job using Skype. Eleven of the 39 participants had been interviewed for a job using Skype. Others reflect on how they might feel about that as a future prospect. Just two of the participants had interviewed others for a position. Of the 39 participants, 21 of the 39 had used Skype for the purposes of their employment. The chapter begins by canvasing some of the experiences people have had being interviewed for jobs on Skype.

Job interviews

Margie, aged in her mid- to late fifties, explains that when she last did a job interview her prospective employer conducted a pre-interview by Skype before deciding whether or not pay the costs of bringing her and a couple of other short-listed applicants to Wellington. When Margie was asked about her experience of being interviewed using Skype she replies:

> Not very good because there was no relationship that had been established beforehand; and so you're with unknown people. And also ... the line [connection] wasn't very good and so the picture kept breaking up. We persisted with it for a while and so that was a bit disconcerting.

Margie faced the strangers on the screen, 'on line', and attempted to 'align' with or 'orientate' herself towards them (Ahmed 2006) in an attempt to secure the job but she could not see them properly and consequently felt disconcerted and thrown off.

Natalie, a university lecturer aged in her sixties, was also disconcerted by not being able to see others properly on the screen and feared the technology failing her in a Skype interview. Natalie says being interviewed using Skype would not be her first choice. She adds: 'I would just be so nervous ... I was interviewed via phone conference call once. We tried Skype. It didn't work.' She adds that being interviewed via Skype would be 'strange' particularly if she was at home: 'Having a job interview while sitting at home but being dressed up, it might be a bit weird.' Krittane, a despatcher in her early thirties, also describes having a job interview using Skype as 'weird' saying 'I don't like it ... if you just sit at home dressed up, it's weird.' Krittane's and Natalie's comments about job interviews via Skype being 'weird' and 'strange' call to mind Ahmed's comments about queering, disorientation and disturbance. People have grown increasingly used to the boundaries between home and work being blurred as they engage with work emails at home, check Facebook posts at work and so on but to dress up at home for a job interview still feels to some a little strange.

Debbie also found being interviewed by Skype to be strange or disorientating. Like Margie she had a pre- or initial job interview via Skype. Not quite knowing how to best approach this, Debbie sought some tips from YouTube, as she explains:

DEBBIE: There were some tips out on YouTube ... so I found whatever it was and it was a guy who dresses in this suit to give you tips on interviewing via Skype. Then at the very end, 'cause it is quite comic, it's very good advice but it's quite comical at the end, he shows he hasn't got proper work pants on. I really got a lot out of that … I was in this room and I turned that desktop computer around and I sat against here and I had the light facing me and I dressed up and I actually even had new shoes on which is ridiculous but there you are. I made a real effort.

INTERVIEWER: To make you feel better?

DEBBIE: Yes, yes, I made a real effort and it went incredibly well because I thought about looking at the camera, I thought about speaking out loudly, directly, smiling which one of my referees had said to do, 'So make sure you smile 'cause it's Skype.' I had a really good feeling about it and it actually did go really well, I realised later. What was ironic about that was that the people at the other end were in this tiny, dingy, little university room and I could barely see them. They were really distant from the camera, it didn't matter at the time but it was just an interesting observation that I'd gone to all this effort to present myself really well and I think I came across really well but, really, not the others.

INTERVIEWER: They hadn't thought it through at all?

DEBBIE: No. I think that was really instructive actually about using Skype in a professional setting.

Debbie's reflections on the job interview, both her own response to it of actively thinking through how to present her body and space (note she chose to locate herself in her office at her existing workplace rather than at home), and her prospective employer's response to it of not thinking through the medium, illustrate that Skype had the effect of creating a 'disorientation in how things are arranged' (Ahmed 2006, 162). Things in a face-to-face interview that might be seen and appreciated such as the interviewee's new shoes and the expressions on the faces of those conducting the interview can be obscured from view. Particular things and bodies through Skype became oblique.

Given how disorientating all this can feel it is perhaps not surprising that a Google search for 'job interviews on Skype' turn up a long list of results (more than 2 million), many on how to conduct a Skype interview, or to use Ahmed's terms, to stay 'in line', 'orientated' and effectively extend the body into the virtual space. A few examples of these sites are: '7 tips to nail a Skype interview' (Forbes), '5 tips for acing your Skype job interview' (Job-Hunt. org), 'Top tips for Skype interviews' (Guardian), 'What not to do during a Skype interview' (Businessweek) and 'The new secrets to rocking your Skype interview' (The Muse).

Some of the advice is very detailed. For example, *The Muse* in a section on 'Dress to impress (not distract)' advises: 'While you want to dress professionally (again, from head to toe, not head to waist), don't just pull out any old interview outfit – take care to make sure what you're wearing works for video' (Dockweiler 2016, no page number). More specific advice follows about particular colours that look 'great' on video (various shades of blue) and particular colours that look 'too bright' (various shades of red). Advice is then directed specifically to 'the ladies', explaining that jewellery could be distracting and risks looking 'sparkly like a disco ball' (ibid.) when on screen. 'Ladies' are also advised that showing too much flesh or skin, for example a low-cut top which might be acceptable in person, might seem inappropriate on screen when one is visible only from the chest up.

What this long list of sites offering highly detailed advice indicates is that many businesses and organizations are now using Skype to conduct interviews. What is more, it is not necessarily 'straight-forward' contending with this new platform. And it would seem from the website above, especially if you are 'a lady' (the advice being not too much skin or too much jewellery), gendered expectations or norms of bodily (re)presentation co-exist in both offline and online space.

As participants in this study indicate, interviewing for a job using Skype can be challenging. It can be difficult to get a feel for the space, to ensure the background is 'appropriate', to create the right affect, to read the bodies of those involved and to know how to dress. Geron told me that she had an interview by Skype once and arranged the shelves behind her 'a little bit so it looked nicer ... took things out that shouldn't be there'. She says: 'You can point the camera where you want to in those kinds of interviews, so that's good; you can choose your own environment which is one of the positives about it. It worked because they offered me the job.' Geron, a lecturer in her mid-forties, had worked out how to do it.

Unlike Geron, Katie, aged in her early twenties, who is employed as an 'operations assistant', feels that trying to present the self and the space projected by the webcam for a job interview would be a near on impossible challenge. She had never done it and is not really keen to try although would if there was no alternative. Katie thinks that there would be:

> so much pressure to look into the camera and to sit up straight and to have your paperwork in order, to not make too much background noise, to have a quiet place. I feel like it would just be too hard in terms of making a good impression and getting the right connection and background noise and balancing the visual and the sound and your first impression of someone. If I had to I would do it. If it was a job I really wanted then obviously I'm not gonna turn down the Skype interview but I would feel really uncomfortable and just feel really dorky [foolish or stupid] doing it.

Skype began in 2003 as a platform mainly to help family, friends and loved ones stay connected over physical distances. As Skype state on their website their aim is: 'to reach out to family and friends worldwide'. Using it for job interviews is a new development. It is a reorientation of the technology towards a cohort of people who often do not know each other and to a different space, the workplace. This can be disorientating for a variety of reasons, one of which is that when Skype is used for job interviews it highlights the merging of the private and the public. Skype has the potential to reveal within its frame the home, car, café or whatever space is used for the Skype call at a time when we are attempting to be at our most professional.

As feminist geographers (e.g. Blunt 2005, Blunt and Dowling 2006; Domosh 1998) have argued over the past four decades the spaces of home have long been associated with women, children, femininity, unpaid labour, and violence but they have also been a refuge for some including black women in a world dominated by whiteness (hooks 1991) and lesbians in a world dominated by heterosexuality (Johnston and Valentine 1995). As Ahmed (2006, 31) points out in her discussion of tables 'To sustain an orientation toward the writing table might depend on such [domestic] work, while it erases the signs of that work, as signs of dependence.' While we might want our future employer to see us as enthusiastic and well-dressed for the occasion, we do not want them to see the breakfast that we prepared ahead of the interview or the wardrobe from where we chose our outfit. Therefore, we exercise vigilance to remove these signs, traces and frames from view. They potentially reveal us as weak, feminised, embodied and dependent. The domestic labour of preparing ourselves for interview needs to be concealed. If it is exposed we, and others, may be prompted to feel uncomfortable and disorientated – or 'weird' as Natalie says.

In order to reduce feelings of 'weirdness' Bailo (2013) in *The Essential Digital Interview Handbook* provides advice on how to create a 'Skype Studio' for people who might be doing a lot of interviewing or who want to make a good impression at high-stakes meetings. It seems that the implicit aim of this is to help avoid revealing too much about one's domestic space. Bailo advises hanging from the ceiling a roll of 'professional background paper in a nice neutral color that does not compete with your wardrobe'. Having a clean, simple backdrop, he says, looks professional and ensures that the interviewer focuses on you. In this case it seems the background needs to be removed entirely to eliminate any risk of it inadvertently becoming highlighted for the interviewer. Next Bailo recommends 'soft, natural lighting' in order to create an inviting space. Bailo also suggests that high-quality sound equipment, a high definition web camera and upgrading the mic are a good idea. Sound levels, how soft or loud we speak, and our ability to hear others speak, are not always easy to manage.

When Haylee, a university lecturer, was asked at the end of her interview if she had anything to add she said:

> One of the things that I find quite funny about the two colleagues of mine who use Skype a lot more than I do and they don't use it in supervision, they use it in co-writing with people, other academics, they yell. They are two older guys and it's like they're just not aware that the microphone's right there. They get louder and louder and louder. It cracks me up [she laughs] every time.

Apart from managing the space, including the oral and aural dimensions of the space, one also has to manage bodily gestures. Again, there is plenty of online advice on this. 'As you're communicating, lean forward,' suggests Bailo (2013, page number unknown). 'This will show interest and concern and will engage your audience. It will also convey eagerness and willingness to listen.' The advice appears to be to not move too little (or the interviewer might think you have passed away!) or too much (or you will appear blurry and none of your facial expressions will be visible) (StyleBistro 2013).

There may also be technical issues to manage. Ewan, a government employee, explains: 'At work we really have problems with the system dropping out; and again it's getting slightly better as the years go by, but it's notoriously unreliable, and there's problems with booking.' Ewan says that he has lost count of the times that he has ended up resorting to just using the telephone because it is too hard to keep pursuing an audio-visual link. Also in relation to technical difficulties, Manee talks about a Skype interview in which her webcam was working but the other person's was not. This is reasonably common and again can prompt interactions which may seem unfamiliar or queer. One party is visible, the other is not. If one is expecting that both parties will be equally visible then this may seem 'awkward' and uncomfortable. Manee explains:

> One time I was being interviewed by a researcher ... we did it through Skype. He was at his home; I was at work ... but Skype wasn't working very well at the time and I couldn't get his picture to come on. He could see me, but I couldn't see him and it was really awkward and I couldn't get someone to come and fix it. So in the end we decided to just go with it.

Sometimes 'just going with it', treading a new path, seems like the best way. Things do not feel quite right or 'normal' and so we 'go with' the 'weird', 'disconcerting' and 'awkward' – all adjectives used within the interview transcripts in relation to work.

Beverly, aged in her mid-twenties, reiterates this point saying that although she would not be 'freaked out' by having a Skype interview because it is a 'serious' activity she would prefer to do it in a workplace face-to-face rather than from home using Skype – it would add stress she says.

BEVERLY: There's something about being physically present there [in a workplace for a job interview]. I think that makes it feel like an event that you have to prepare for and you have all the 'faff' [getting dressed, prepared] around it where you go in and you're wearing your interview suit and you're standing there physically and you wait outside ... things like just the act of walking into a room puts you in an interview mode I guess. So being taken out of the space kind of makes the whole interview process seem less... I'm trying to think of the word and I can't. I know what I'm trying to say!

INTERVIEWER: Less like an event?
BEVERLY: Yeah. In some ways, too less serious.

Beverly makes the point that not being interviewed in the physical space of the potentially new workplace means that candidates are not provided with an opportunity to see and feel whether it is somewhere they would like to work. Kurt, aged in his early sixties, also notes this: 'If I was being interviewed for a job somewhere and it was done online, on the, you know, Skype, it means I wouldn't be able to get to the place to check it out, to have a wee nosy around and get a feel for it.' When Skype is used for interviewing prospective employees the workplace itself is relegated to the background or maybe completely absent from view. The interviewee's attention is directed away from it and placed instead more squarely on the people interviewing. Skype offers, therefore, a different view than an interview that takes place offline in a real workspace. In this way, again to return to Ahmed (2006), bodies become (re)orientated, taking up time and space in a different kind of way. We are used to performing our bodies, repeatedly (Butler 1990) in particular kinds of way (such as wearing pants for a job interview). When we enact different performances it can have the effect of queering the bodies and the space.

Travis, a participant who is retired, mainly uses Skype to stay connected with family, including his granddaughter, but I asked him to reflect on using it for a job interview. He thinks for a moment and then replies:

> You're not interacting the same [as face-to-face] … but they're [job interviews] probably are a bit stiff and formal anyway because you're a bit scared about saying too much or not saying enough. I think the same would apply; you just wouldn't be able to read the body language. It's probably not the same, you're probably not as relaxed anyway with a job interview but I would say you would probably be less relaxed if you're doing it by Skype. Once again the only advantage I suppose is you don't have to wear a suit, you can just wear a shirt, tie and shorts if you want to 'cause no one's going to see the rest … and props [notes to help answer questions] all set up around you.

Again, we see in this exchange how bodies, objects and spaces – clothes and props – are reconfigured through Skype for the job interview, or at least how Travis imagines they might be. Travis assumes that not having to wear long pants is beneficial. Others, however, comment that they would find this unsettling or disorientating because it does not feel 'normal' or 'natural' to do a job interview in casual attire, or half in formal and half in casual attire even though the casual half of the body may not be seen (see Figure 6.1). Wearing no pants for a Skype interview has become something of a joke in popular culture. A Google search for 'no pants for Skype interview' reveals more than 80,000 hits including numerous memes, images and videos.

Figure 6.1 Skype job interview: jacket and tie but no pants

Brittney, a student aged in her early twenties has not been interviewed for a job via Skype but says that she thinks it would make her feel 'uncomfortable'. She relays a story about her father having recently been interviewed for a job by Skype. Brittney says: 'He [Dad] had to borrow my laptop and find a place to do it. He was in this café and he had his phone with him as well and he had it all set up on the laptop but it came through on his phone and then he ended up doing his Skype interview in his car.'

These reconfigurations of bodies, objects and spaces (e.g. clothing, laptops, phones, offices, cafés and cars) were not necessarily an uncomfortable experience for *all* of the participants though. Nigel, aged in his late twenties, had never done an interview using Skype but comments: 'I don't have a problem with it' Quentin, aged in his early forties, goes a step further suggesting: 'In some cases it's better for job interviews maybe … nobody knows who you are behind a computer screen. Nobody knows you're not wearing pants behind the computer screen. All they see is the t-shirt or tie.' Quentin is quite comfortable being able to craft a particular version of himself on screen. Somewhat contradictorily though he also mentions that when he was once offered a job interview he was somewhat disappointed that his prospective employers decided to *not* fly him to another city in Aotearoa New Zealand's South Island (which he said would have made him feel 'very flash') but instead did it on Skype. While Quentin thought this situation of doing the Skype interview was not ideal, it 'was okay'. He explains:

> I kind of feel like job interviews don't need to be done in person. They can be done via Skype in that it's a waste of … it's a carbon foot print being flown somewhere, especially to another country as well. That's something where Skype can fill that void, or video. Not necessarily Skype but any kind of video, computer to computer interface. But as far as the actual interview was concerned, I don't feel as nervous I don't think. There were nerves but I was in a controlled … I was in my own space when it went down, when the interview was conducted, so I had control over my immediate area, and I didn't have them right in my … you know, I wasn't physically present, so I guess that made me feel a little bit more relaxed. Although I was concerned that the technology could fall over, which is why I didn't do it from my [previous] home internet connection … So I did it from my office [individual office Quentin had at university as a PhD student] and it was rock solid because they were doing it from their office as well. But yeah it made me feel a bit more confident. It just ruled out one of the things that could throw a spanner in the interview works.

Having 'control' over one's own 'immediate area' might include having notes available – perhaps a nearby wall – which the interviewers are unaware of.

Denise too had been interviewed by Skype and said it did not 'phase' her. She had thought initially that she would be 'really nervous' but was actually 'really excited'. She describes herself as a 'communications person' and as being able to make the interview less disorientating by visualizing that she was actually there [with the people interviewing her] 'in order to make it more natural'. Denise recognizes that for many being interviewed for a job by Skype might actually prove challenging but she was able to make it work for her.

Imogen is also an effective communicator and as a digital artist is an early adopter of new technologies. She began using Skype when it was first released in 2003 and has been interviewed 'a few times' for a number of positions via this medium. Despite feeling very comfortable with Skype, interviews via this medium are not always easy partly because of the time and space differences it can prompt when Skyping between countries. This can prompt bodies to feel out of place. Izzie, who is currently living in Europe, explains:

> The last one that I did was actually when I was back in New Zealand over the last Christmas and we were staying at a holiday house out at [name of small settlement]. I applied for a job, it was a project in the UK and they wanted to interview me by Skype and the house where we were

staying had internet but it was terribly bad, so I arranged to go into [name of city] my sister's house – my family was all staying at the holiday house but only 20 minutes away from [name of city]. The interview was for 11 pm at night New Zealand time which is 9 am in the UK so we couldn't really do it; I mean a few people had to get together in the UK for the interview … it was two nights before we were having my Mum's 80th birthday; we were all quite busy getting ready for that. [Name of partner] came with me; we went into my sister's house in [name of city] and got there and the internet wouldn't work and I couldn't find the modem. I spent ages looking around the house for the modem and then I had to call my sister and luckily she had her cell phone on and told me where the modem was and she said: 'This happens sometimes you just need to restart the modem,' which was what I was expecting; why I was looking for the modem. So I restarted it and it worked and fortunately I had got to her house early enough to be able to do that. So at 11 o'clock I was online but I was really tired and I didn't feel like I looked very good. I just didn't feel great. I felt tired and because we'd had a little bit of stress with the modem first. The Skype interview, there were four people there gathered around the camera at their end and I got introduced to them all. Their computer kept being slightly out so there was nearly always two people that I couldn't see properly, which didn't happen too much but we spent quite a lot of time trying to fix that and they were moving the computer around. So there was a bit of distraction with the technical stuff, which was okay because for me I've done it before. Actually it was a pretty good interview, like I woke up a bit once I started talking and I got the job … At the end of the interview I just thought, 'Oh thank God that's over;' it was really quite a mission to do it at that time of night and when I was in the middle of a holiday and that sort of thing.

This story is relayed in detail because it encapsulates so much about the Skype experience that many participants described. Not everyone has had to manage time differences but the feelings of bodies on Skype not quite being in the right place whether it be being at one's sister's house and not knowing where the modem is or a number of people not being able to fit comfortably on the screen together, there is a sense that things are not quite as they usually are.

Bodies and objects are not quite lined up or at least it is a different line-up across time and space which can cause stress, despite users being technologically proficient, because we are not sure where this line-up might take us. Ahmed (2006, 7) drawing on the work of philosopher Immanuel Kant in his essay 'What does it mean to orient oneself in thought?' asks readers to consider the experience of walking blindfolded into a room in order to think more about orientation. Ahmed says that finding one's way depends on knowing the difference between left and right, knowing how to turn. When using Skype sometimes we do not quite know how to find ourselves in the room, which way to turn. Izzie faced four people in the room all gathered around the computer – sometimes not all were able to fit in the frame – they were 'slightly out' – the computer had to be moved around so that everyone could fit or be orientated correctly.

While a total of 11 participants had used Skype to be interviewed for a job only two had interviewed someone else using Skype. One of these is Ewan, a government employee who has been on interview panels, usually with a minimum of three people. I asked Ewan if he is able to get a good sense of the person through the screen, to which he replied emphatically

EWAN: No, no you don't get anything like a good sense. It tends to be more formal; there are nuances of their body language, of their interaction, their ability to engage with you as a person. So what's going on with the eyes, get a sense of how they're going to be in difficult situations where they have to do a facilitation of complex challenges where there's high emotions and so on.

INTERVIEWER: You can't get that via video conferencing?

EWAN: No. I feel you don't, you don't tend to get that in the same way.

INTERVIEWER: What view of them do you get? Are they sitting?

EWAN: Yeah they're always sitting; and sometimes they're quite distant. Sometimes also the thing will freeze on you and you'll get a disjuncture between the picture and the words. It's not crisp and clear.

Ewan's use of the word 'disjuncture' for me immediately conjured up Ahmed's (2006) work on disjuncture, discomfort, disorientation and distance. In fact, in relation to distance, the conversation continues with the question

INTERVIEWER: So you said you don't get to see their eyes? Is that because they're too far away from the screen?

EWAN: Yes. And even if you did you don't see enough of it to really get a sense ... it's like all the creases are blurred out. There's no crinkle of the smile ... They're [applicants] much harder to read [on screen]. It gets lost; the feel, it gets lost.

It is useful here to return to Ahmed's (2006, 6) point that: 'When we are orientated, we might not even notice that we are orientated: we might not even think "to think" about this point.' During a job interview, both for the interviewee and interviewer the creases of a face and the crinkle of a smile, are likely to be taken for granted. There are bodily cues that direct us towards how to think, act and feel. It is not necessarily until we are without these bodily cues that we notice them as missing and therefore may feel disorientated.

Meetings and collegial communications

This section builds on the participants' experiences of job interviews by looking more widely at their experiences of using Skype in workplaces.

Sometimes people use Skype at work to meet new contacts, students and colleagues, other times it is to meet with an employer or colleague whom we already know.

Scott Dockweiler from *The Daily Muse* offers careers advice including on meeting new people. He says that given how much people's contact at work is through synchronous audio-visual it is important to 'master the digital handshake' because, as with in-person encounters, first impressions matter. Dockweiler notes that in the first few seconds a 'digital chemistry' is created. And yet terms such as 'digital handshake' and 'digital chemistry' are still not common for many of us. We might have a sense of what is being talked about but not really feel that we would know how to do this. Dockweiler (2016, no page number) explains:

> a 'digital handshake' involves a: 'slow, confident, professional, firm nod' with 'a slight shoulder bend and eyes forward – the other person should not see the top of your head.' [Bailo, 2013] ... From then on, focus on keeping your eyes on the camera – not on the view from your screen. 'Your eyes need to look straight into the camera, so it appears on the other end you are looking right at the other person,' says [Paul] Bailo.

Not everyone feels comfortable with this new orientation. For example, Barbara, aged in her fifties, who had worked for an employment recruitment company explained that her manager insisted employees remain logged onto Skype all day in case she wanted to contact them. She was asked what that was like, to which she replied:

> Really annoying because she could see ... we always had to be online and stuff and so she'd just quickly send us a message or wanna Skype all the time. It was really annoying. I hated it ... I found it intrusive, yeah. ... If we weren't online and she wanted to Skype us she'd just either text us or ring us, get us to go online so she could Skype us. So yeah, that was annoying; intrusive.

It is worth noting here that Barbara's employer did at least send her an instant message over Skype (which some describe as good Skype etiquette – the equivalent of knocking on the door of a real office) to check availability before actually calling but still Barbara found this to be intrusive.

Desiree also uses the word 'intrusive' to describe a Skype call from a colleague in Texas one evening:

> This colleague in Texas calls me sometimes and when my iPad at home went off at a time when I didn't expect it ... I was like, 'Oh,' and I open it up and you're there, I suppose that maybe felt a little intrusive. That has probably more to do with the sort of home/work boundary; it was like I don't want to do this right now.

Ewan works for a government body that manages school properties. I asked if he and his colleagues are more likely to feel under surveillance if in the future they had to check in with line managers via audio visual link when they drive out to rural schools. He says probably not:

> but what it might do is make people in schools feel more under surveillance. Certainly there is a hierarchical relationship that the Ministry of Education are a fund holder, and schools deliver something for them; the person who holds the purse strings undoubtedly has more power. So they may well feel, you know if we're asking to see something around the school, well we can make a judgment, in fact, the Ministry person might not think anything of it but it's always different on the other side, the less powerful. So they might feel under surveillance.

Ewan says he has spoken with a number of his colleagues about using Skype to meet with staff in schools and for them to walk around properties with a laptop with Skype to show Ministry staff. Ewan, however, explains:

> There's a strong feeling that the face to face communication is the corner-stone of what we do, of establishing a relationship; and we don't want that to be lost. We know it's not always possible but it could be decided this [using Skype] was a cheap way, a way to cut costs and increase output; it would be a very poor quality output.

Margie who has in the past worked as a school principal, found Skype to be useful because she was in a region that is quite geographically isolated. A principals' professional support group was set up and they met using Skype. Margie said that she had already met all the people in 'the [virtual] room' before face-to-face. She says 'it's good to have the picture; but I think if the picture is not good ... I prefer just to have audio because then you can just focus on the audio ... rather than being distracted by a picture that's quite poor.' Margie was not the only participant to say this. It seems that on account of the long history of phones people tend to be well accustomed to hearing a voice without an image and so this is often less disorientating than hearing a voice accompanied by a blurry, stilted or pixelated image.

Francine also used to work in a school but as a teacher at a preparatory boarding boys school (for children aged 7–13 years in the United Kingdom). She explains: 'We have a lot of overseas students and they access Skype weekly to contact their parents in Spain or Korea and it's great for them as well to see their parents and their family and get passed around ... the parents really enjoyed it if you popped over the kid's shoulder and say hi as well.' Francine was the only school teacher to be interviewed and so I do not discuss Skype in relation to schools extensively in this chapter on work but it is worth pausing just a moment to at least signal that Skype has become a routine part of school work, including teachers discussing material with students, students working with each other locally and making contact with other students in other places including overseas.

When I asked Geron about whether she uses Skype at work she said that she was unable to attend the staff meeting scheduled to take place the morning after our interview and so she asked her line manager if she could join the meeting using Skype. Geron notes that she will be the first to do this and so feels rather like 'a guinea pig' trying it out to see if it might work so that others in the future can also do it.

Haylee, a university lecturer, sometimes uses Skype for supervising graduate students. With one particular student who has a disability she finds Skype to be very helpful. Rather than disorientate the able body that usually fits comfortably into offline space, in this instance Skype helps orientate the body through Skype (see Davidson 2008 on virtual communication enabling people on the spectrum). Haylee explains:

> One of my students is seriously disabled and part of her disability is that she can only whisper and she knows that it's really hard to hear her and so when she was having some really serious problems she was like, I will Skype you. So we ended up Skyping 'cause she's auto-immune suppressed as well, so it's bad for her to be around germs and we just had the best conversation because I think she might be a gamer because she's got a great Skype name ... She just was so set up, she had a little headpiece and a microphone so I could actually hear her and we had a normal conversation, it was so nice. I just feel like crap when she's desperately trying to communicate with me but there's other people talking and I'm hard of hearing ... when someone can only whisper it's just like 'Oh my God.'

On the theme of disability, Haylee added to this that one of her closest sisters has been suffering long-term with glandular fever and it is useful to be able to Skype each other 'because she may be on the couch that day with a mild fever and it just saps all her energy'. It is important to remember that not all bodies are the same and a communication technology that may disorientate one body may orientate another (Moss and Dyck 1996, 2002).

Howard, aged in his late thirties, works on a research station on an island in the Solomon Islands. The island is 5 kilometres from the mainland and he is one of just three who works there. Howard's head office is in Malaysia and his partner lives in the city of Honiara, the capital. Skype is important to Howard who uses it on a daily basis to 'look after farms', to help decide where to use money for projects and to correspond with scientists who live on other islands. Howard says that it is a cheaper method of talking with people than phoning which is expensive. He does not always use Skype with video though. Sometimes just the voice is adequate and provides him with a quicker response than texting or emailing. Howard says that without video: 'I can just feel good ... sit back and don't look at anything, and express myself and maybe walking around … it's like if the visual is there it's almost like you have to think about how you're appearing don't you?' Also, Howard works in an open plan office and so it is not always easy to control what might be happening in the background plus the call with video might be distracting for others also in the space.

In relation to what is seen in the background when using Skype at work Ewan says: 'It can be quite distracting particularly when there are people using offices with glass partition backgrounds, and then people are coming up and looking in. You can't hear them but you know what's going on, and you get distracted by that.' He also finds the image of himself (refer to Chapter 4) to be distracting partly because he thinks he looks 'awful' but also because there is a slight delay in the image that he sees of himself on the screen. Ewan says: 'and it can be hard to feel confident about that awful looking person there, or this kind of weird – you move and then of course a moment later it moves! Disconnected … I'm more self-conscious [on audio-visual]; yes, absolutely. More stilted.'

Olivia often has board meetings with iwi [tribal] members when she gets home from work at night starting around 7 pm since this is when people are available. She is very aware of both what she is wearing and the background in thinking about what the camera might capture. Olivia explains:

> So I'll be in my track pants but I'll keep my work kind of attire from shoulders up, make sure my hair is done. And I think too 'cause you can see yourself back you want to probably, even more so, maybe scrutinize what you look like. And definitely make sure that whatever the background is that I leave it plain, very plain, can't see that you've got dishes on the bench. I tend to keep it pretty plain. Even at work if I Skype at work I'll kind of make sure that they can't see the chaos behind me or even my whiteboard that has a whole lot of random messy notes to myself on there. I tend to face the other way so that there's a bookshelf behind me rather than my own work behind me … And the other thing is sometimes if I have a board meeting at night and I haven't had any dinner. I'll often have my dinner next to me and kind of just mute the microphone and eat while the meeting is going on. I find that sometimes; it actually lets you do a couple of things at once.

INTERVIEWER: So you're off the camera while you're eating as well?

OLIVIA: It depends what it is; it depends what the food is or I'll make myself a quick sandwich.

INTERVIEWER: Not a big burger?

OLIVIA: Not a big burger. Might get a mandarin or something; probably more inclined to do that, usually 'cause it's in the evening type stuff.

INTERVIEWER: It's interesting, you've kind of also got to think about what sort of food you can eat while you're having a Skype conversation, 'cause you can't just eat any old thing like a burger?

OLIVIA: No, or like a thing of spaghetti bolognaise where you'd slurp in some noodles. No, so I'll usually just grab something that I can kind of quickly pop in my mouth and if I need to un-mute and add my piece.

Skype, however, involves not only consideration of the space occupied, backgrounds and foregrounds, eating and drinking and if acceptable what kinds of food and beverage but also if one is Skyping overseas, time differences. Howard makes the point that in relation to colleagues at work:

> Normally what we do is we email each other prior to Skyping so that we are aware and get the time straight. So like for example Solomon Islands is three hours ahead of Malaysia in Penang which is our Head Office. So every time normally they give us a time in Solomon Islands because they are still sleeping when we start work kind of thing.

While many of the participants in this study who are involved in education or research sectors use Skype as part of their working lives, others involved in trades tend not to. For example, Harrold, who is a boiler maker and welder, says that he has never used Skype at or for work but his 'bosses use it all the time, for conferences, with other sites, other companies, yeah they're a big organisation, big tool organisation, they use it all the time. They probably wouldn't survive in this environment without it.'

Imogen would also not 'survive' without Skype and other platforms for streaming and real time interactions. As mentioned earlier she has been using it since it was first launched and has become adept and at ease with the technology. She says:

> I work with a lot of people who are not in the same place as me so I use Skype a lot for project management, planning, meetings; everything for my work and also in performance. I've been working a lot not so much with Skype but rather with streaming in various different ways as well as other kinds of real time interactions through the internet. So that's how I use it in my work, pretty much in every bit of it really.

Imogen mentions having organized conferences, worked collaboratively with other artists, editing a book with someone interacting on Skype daily, carrying out online project management in the internet industry and performing where people and audiences are online in different places and at different times. She says she has been doing these kinds of things since 1999 'even before we were using audio visual with a bit of graphical animation of things'. Unlike most of the other participants in this study Imogen appears to have become thoroughly orientated to life via screens. Offline and online blur for Imogen. Skype is not a new medium for her.

Miller and Sinanan (2014, 186) make the point 'that people at first focus on new technology to overcome technical limitations and frustrations within the status quo, and only later on look at less precedented possibilities'. Imogen does not only use Skype to solve the problem of how to connect with family back in New Zealand but uses it creatively as a writer, theatre practitioner and digital artist. She appears completely oriented and able to find her way in a room that others experience as dark (Ahmed 2006). Miller and Sinanan (2014) explain that when people first use a new communication medium they tend to be self-conscious. While this was the case for many of the other participants in this study, it was not the case for Imogen. She comments:

> I'm working on editing a book with someone who's in France and we talk every day on Skype and sometimes we have Skype open while we're actually just working so that we can kind of mention things to each other ... sometimes I almost forget; like I've just come back and I'm in my yoga clothes, my hair might be messy or I might have just got out of bed and I look like crap and I don't really think about it too much.

James, a university lecturer, raises the point that Skyping one person is different from Skyping a number of people. When meeting with a number of people, even without video, the space can become disorientating by way of not quite knowing who ought to speak when. Without clear visual clues this can prove challenging. James also notes this explaining:

> It's not too bad when it's one-on-one. I Skype colleagues in Manchester sometimes ... that's fine if it's just one of you. As soon as you move up in scale it becomes a very different beast. You lose the quality, it becomes more reporting in like a traditional meeting kind of thing. You don't know when to start talking or when you don't and it's always everyone talks and then everyone shuts up. Then everyone goes, 'No you go' at the same time. It's all that business ... it's a different way of speaking, you have to shut up, you don't do the 'yeah', the reinforcement signals because if everyone did it would just be like, 'What did you say?' 'Nothing, I was just saying, yeah, go on.'

As mentioned previously Microsoft Office in 2015 launched a new mobile app across Windows, iOS and Android called Skype for Business that replaced earlier versions of communications software offering a combined platform for calling, conferencing, videoing and sharing. Like Lync and Skype Premium before it, Skype for Business allows for multiple presences. This will likely in the future be popular with many organizations, institutions and businesses given the many committee and board meetings held on a daily basis on audio-visual platforms. Microsoft claims that it is the answer to making virtual meetings easier and more collaborative. Many participants in this study, however, were not overly keen on meetings with multiple people via Skype or any other platform.

Debbie attends editorial board meetings online. Some members are in a room physically together while others attend virtually (audio and visually). Debbie did not enjoy the last one.

> At the last meeting I think I might have been a little bit too blasé because, not that I was doing anything untoward, but I got so bored with it that I was, I was kinda stretching back. I was rolling around like this [lays back and rolls to one side in her office chair]. I was extremely irritated by the whole meeting ... I actually did think later, 'Oh God I wondered if people noticed how really irritable I was?' It had become almost like I was invisible. That was my emotional reaction to the meeting, it was how I behaved, I was invisible but actually maybe I really wasn't.

One of the things that Debbie's narrative highlights is that Skype is about what we see on the screen but it is also about others seeing us. She felt that maybe others were not seeing her, that she was invisible (Debbie was feeling ignored because the others were all in a physical space together) but realized later when she thought about it, that in fact, they probably could see her quite clearly.

In discussing using Skype at work Miller and Sinanan (2014) remind readers that Skype is not the only communication media used. Webcam may come through Facebook, Yahoo, Googlechat, FaceTime or MSN. People will also, and sometimes simultaneously, message, voice message, text and email colleagues. They might also be in contact with friends and family at work. Sometimes people just have one monitor but a split screen, others have several screens. Madianou and Miller (2012, 3) define these constellations of different media as 'polymedia' pointing out that it is useful to shift attention 'from the individual technical propensities of any particular medium to an acknowledgement that most people use a constellation of different media as an integrated environment in which each medium finds its niche in relation to others' (also see Nayar 2010).

This is not entirely new of course. People in the past sent letters, cassettes with recorded messages and phoned but today there are many more opportunities for communication. In the conversation above with Debbie she describes feeling bored with the meeting she attended. She does not identify in detail the source of her boredom but it may be that she wanted to be involved simultaneously in other activities but because she was on screen this was not possible and she felt as though she had to stay focused on that one activity.

Olivia, while recognising the value of Skype meetings for bringing people from a range of different geographical locations together, also has some reservations. She is a board member of a charity organization, Chair of a Māori organization with members located across the countryside. Olivia says:

> Predominantly my use of Skype has been for meetings ... So for the [name of iwi] Board that I'm on which is an organisation that works with

marae to reduce their waste – there's staff all across the North Island and down to Nelson as well at all different places. Anyway in May we had our big yearly workshop and so that was at the marae, so that was an in-person one and the Board for the first time met in person at that workshop, and so that was quite interesting. We got together in person and we actually started our meeting talking about how nice it was to be in the same room face-to-face. We kind of raised some of those things that we were talking about, that it can be quite hard to read other people. Often on Skype and on teleconferencing, I don't know if it's a technology thing, but I think you wait until someone said their whole piece; it's kind of like taking turns I think on Skype and teleconferencing whereas when you're in a face-to-face situation it might be more of a dialogue happy to kind of jump in over somebody. So I think we were kind of discussing that, how it does change the dynamic a little bit of the group.

Olivia reiterates a comment made by James earlier. The flow of conversation is different via Skype. It can be hard to know when to come into the conversation. As is suggested in Chapter 3 on using Skype to conduct some of the interviews for this research project, a hongi [Māori greeting in which people press noses together], touch on the shoulder, shake of the hand, or kiss on the cheek can show that the role of bodily rhythm in constituting the social world ought not to be underestimated. Conversations via Skype do not necessarily flow in the same way via Skype with video. For example, Olivia says: 'There was one time where I didn't realise that I had dropped out of the conversation and I was talking for a good few minutes thinking no one agrees with me 'cause it's like dead silence, but there was no one on the other end; I'd kind of fallen out of that conversation.'

Olivia continues:

> There's one guy who hasn't been able to get onto Skype so he kind of teleconferences into our Skype meeting and so we'd never met him in person; so that was quite nice to meet him face-to-face. So we did have a bit of a talk about that but then we equally said it's really convenient. One woman said, 'Oh yeah our first board meeting I was in my pyjamas' and I was like, 'I couldn't tell,' and she goes, 'No you can't.' Our board meetings are like 8.00 till 10.00 in the morning, or 8.00 till 9.30. She was like, 'Yeah that day I didn't have to go to work so I just stayed in my pyjamas.' So we kind of got to chat about the fun side of it as well. Another guy he phones in on his way to work 'cause he doesn't start till 9.00, so he's on the phone in the car. So she's in her pyjamas, he's on the phone in the car and I'm at my desk with my Smart phone. So we kind of have talked about that a bit.

While Olivia talks about some of the disadvantages of connecting virtually, she is also very aware of some of the advantages. One advantage is that because they are a charity organization, if board members meet face-to-face then they have to pay out of their own pockets and, given that they live in three different cities, this can be costly. Another advantage she mentions is that, as chair of a board, at one meeting she was asked to do an opening and closing karakia (Māori incantation used to invoke spiritual guidance and protection and often used to increase the spiritual goodwill of a gathering). Olivia explains that while doing the opening karakia was fine, when they got to the end of the meeting she suddenly had a 'mind-blank' and realized that she had forgotten the words so, because she was on Skype on her phone and had her computer handy, she was able to search the closing karakia using Google, unbeknown to her colleagues.

Given the complexities of conducting a meeting with these multiple body space configurations perhaps it is not surprising that Queenie, who is a Property Advisor for schools, says that she still prefers to actually go and visit schools and have face-to-face board meetings because 'you get more interaction than over Skype.... you're not usually getting a clear message across, you're just chatting and interacting, whereas business meetings are planned and there's an agenda'. She prefers face-to-face despite the fact that visiting a site is far more time consuming. Quentin also makes this point saying:

> It's just so much quicker [to meet via Skype]. A two-hour meeting can take you all day if you've got to fly to Wellington to have it; by the time you get out to the airport and get on the plane and get there, get into town and then have the meeting. A two-hour meeting is your whole day gone.

'Disorientations'

There are many issues that could have been discussed in this chapter in relation to Skype and work. I am thinking here about use of Skype from employers' perspective, what new forms of transnational labour Skype might make possible in the future, what might change if more international call centres begin to offer audio-visual calls rather than voice only calls, and security issues currently being faced by companies and organizations using Skype or similar platforms. For example, Ewan says the government ministry that he works for does not support Skype or other similar platforms because 'the government are concerned that it is "a cyber risk" and they are unable to control where information is going. ... It's just this idea that they can't control where information, especially people's personal information, goes.' He says: 'That's why we can only use an internal system, we can't use Skype.' All of these are interesting and contemporary issues concerning Skype and work but it is not possible to cover them all in this chapter.

Instead the focus here has been on issues of bodily engagement with the medium, which came to fore in interviews when people were asked about using Skype for job interview and what it is like to meet with colleagues using Skype. Perhaps a more fitting quote from Ewan then is when he says:

> There's a tension in my body still when I'm around it [Skype]. I'm not quite sure how to act. I also don't feel confident. I should do. I should feel really confident about technology. I do use it. I use computers at my work every day. I don't know but its somehow [pauses] there's a bunch of things that are more like minor anxieties than they are major obstacles, but that is sufficient just to... it's just so it doesn't feel like it's something that's comfortable.

What transpired in this chapter is, that like Ewan, many participants reported feeling comfortable using Skype at home to connect with family, friends and loved ones but this was not necessarily the case at work. Unlike the professor discussed at the beginning of this book, who used Skype so effectively to take part in a book launch, the research participants tended to find the experience of using Skype at work uncomfortable. While the professor was totally at ease in his workspace, confident in his embodied performance and able to captivate his audience despite not seeing most of the approximately 50 people in the room, many of the research participants instead found the experience to be 'disorientating'. It is not what many of us are used to.

A glimpse of domesticity in the background can queer the work meeting or the job interview, it is possible to wear a jacket on top and no pants on the bottom when talking with a future employer, employers can keep track visually of employees and it can be difficult to know when to speak during a Skype meeting. Sedgwick (1993, 8, italics in original) refers to queer as 'the open mesh of possibilities, gaps, overlaps, dissonances and resonances, lapses and excesses of meaning ... [that] aren't made, (or *can't* be made) to signify monolithically.' While Sedgwick is referring here to gender and sexuality she herself notes that '"queer" can [also] mean something different here' (ibid.). In this instance Skype has the effect of potentially shifting existing subject positions, for example, by blurring public and private personas and spaces. What for years has felt familiar is now, for some at least, feeling a little strange. Bodily schema when filtered through Skype change. Currently in Aotearoa New Zealand Skype is still not entirely embedded as a normal or comfortable part of people's working lives. Nor is it embedded as a normal or comfortable part of most people's romantic and sexual lives.

7   Skype sex
'Queer effects'?

Ahmed (2006, 65), in addressing sexual orientation, begins with a quote from Maurice Merleau-Ponty's *Phenomenology of Perception*:

> If we contrive it so that a subject sees the room in which he is, only through a mirror which reflects it at an angle at 45 to the vertical, the subject at first sees the room 'slantwise'. A man walking about in it seems to lean to one side as he goes. A piece of cardboard falling down the door-frame looks to be falling obliquely. The general effect is 'queer'.

Ahmed explains that Merleau-Ponty is interested, like herself, in how subjects 'straighten' any 'queer effects'. One of the findings from this research was that most of the participants thought that sex via Skype was somehow a queer effect – contrived or unnatural – of the audio-visual that needed straightening. Skype, it seems, mediates an act that most participants think involves an intense, emotional and material connection between real bodies, in offline space and in the process makes it a little strange or queer. Skype for most of those interviewed was only capable of producing an artificial version of the real, or a queer version of the straight. Participants still tended to regard it as a poor substitute for real co-present intimacy despite that many people now meet online and have online relationships. Research on online intimacy has grown rapidly over the past few years as more and more researchers have taken up this topic (Baker 2002, 2005, 2008; Chow-White 2006; Cooper and Sportolari 1997; Lea and Spears 1995; McKenna, Green and Gleason 2002, Wildermuth and Vogl-Bauer 2007).

Most of the participants in this research, however, still regard this as somehow deviant, moving away from the straight and narrow. Miller and Sinanan (2014, 67) make the point that 'Sex has always been in the vanguard of online technological developments; webcam is still associated with the phenomenon of camgirls (Senft 2008), pornography and other ways in which webcam aims masturbation singly or mutually' (also see Del Casino and Brooks 2014 on 'Viagra, YouTube and the public(ized) sexualities'). For some of the participants, intimacy via Skype still held this connotation and yet when people are in long-distance relationships and/or are separated by physical distance then not having sex is likely to become 'an issue' at some point. It would seem that over the past decade, the spread of digital media has reshaped intimate relations worldwide, transforming the ways in which people experience sex, love and romance in their everyday lives and yet for my participants it seemed there is still a great deal of apprehension about using Skype for sexual relations.

This chapter examines people's experiences of sex, love and romance through Skype. Again it reports on findings from the interviews with 39 participants who were asked about their experiences of using Skype including whether they would consider, had in the past, or currently use Skype for sexual encounters. The chapter also reports on text posts (there are more than 22,000) on Reddit located by searching the phrase 'Skype sex'. Data were analysed calling on a number of themes including morality, embodiment, gender, performativity, queer, distance and visuality but also keeping in mind Ahmed's ideas about bodies taking the shape of the spaces they occupy and of the work they do. While it seems as though bodies might be taking the shape of the Skype spaces they occupy with family, friends and loved ones for staying in touch and special occasions they do not appear to be taking the shape of the Skype spaces for sex, love and romance.

As the world becomes more interconnected via digital technologies such as Skype with video there is increasing awareness that some political, economic, social and cultural relations are being reconfigured (Horst and Miller 2012; Shapiro 2010). This is happening not only at the global scale but also at the local scale. People conduct their personal, familial, emotional and affectual lives in a myriad of ways and both offline and online spaces through gendered, sexed and sexual bodies. Bodies and spaces – the organic and inorganic – comingle and this includes our experiences of sex, love and romance (Ben-Ze'ev 2004; Brown 2012; Buchanan and Whitty 2013; Cornwell and Lundgren 2001). As has been seen in previous chapters digital technologies have the potential to change what people do every day, including romantically and sexually, in online and offline spaces. It is important to understand more about this. Kinsley (2013a, 365) argues more 'materially grounded geographical studies of the digital' are overdue and as is evident in this book, as a feminist geographer with a long standing interest in embodiment, I totally agree.

In particular, there is space for more materially grounded studies of sex, love and romance. I do not mean voyeuristic tales of webcam of which there are already a number but stories about relationships and intimacies especially those in long-distance and long term relationships. Research is emerging on topics such as cybersex, cybercheating (Woods 2008), erotic role-playing in online games (Todd 2015), hypersexualisation of avatars (Yee 2014) and 'pornospheres' (McKee, McNair and Watson 2015). There is also quite a substantial body of work on online dating (Frost et al. 2008), in-game marriage and weddings (Wu et al. 2007; Yee 2003) and couples meeting online, but less is known about what sometimes tends to be considered the more mundane sexual and romantic relationships between partners, lovers, boyfriends, girlfriends, husbands and wives who miss each other because they are physically parted or want to try something new as part of their intimate lives together (Baker 2000; Barraket and Henry-Waring 2006).

This chapter focuses on research participants' experiences of sex, love and romance via Skype rather than on sexual encounters or 'hook ups' with strangers in part because none of the participants discussed this and because there already exists a reasonably extensive literature on this (Maginn and Steinmetz 2015). Instead it focuses on participants' sexual, loving and romantic encounters, past and present, with partners, lovers, boyfriends, girlfriends, husbands and wives who use Skype to connect when physically apart. Participants were asked whether they would consider, had in the past, or currently use Skype for sexual and romantic encounters. This question about intimate relations on Skype was one of approximately 10 that participants were asked about their use of Skype more generally. The topic was approached with a high degree of sensitivity. Often the research assistants and I used the term 'intimate' rather than 'sexual' because we sensed participants might be somewhat reticent to talk about their experiences. To remind readers, of the 39 participants, 25 were women, 14 men. Twenty-eight were interviewed individually, the remainder were interviewed as a pair with the exception of one group of three friends who were interviewed together. Eleven of the interviews were conducted via Skype, the remainder in person. The youngest participant was aged 16–19, the oldest 65+.

In order to provide some context for the participants' comments about their sexual experiences (or lack of) on Skype I also sought data from Reddit searching the phrase 'Skype sex'. I did not pay attention to all the posts since numerous entries appear under the heading 'Skype sex'. Instead I focused on posts which addressed themes surfacing in the interviews such as uncertainty about how to be romantic with a partner via Skype. Also, throughout the duration of the project I kept a folder of relevant media, Facebook posts, Tweets and so on that came to my attention (see Döring 2002 on studying online love and cyber-romance). This chapter though reports primarily on the interviews, supported by an examination of these other materials. These data were analysed calling on a number of themes including embodiment, gender, performativity, touch, distance and visuality.

I was keen to ascertain whether Skype is providing a space for expressing a variety of different kinds of love, sex and romance. It may be useful for simulating situations of co-presence (e.g. sharing a meal or mothers and children sharing stories) but is it successful in enabling people to share more intimate sexual and romantic moments? How important is visuality (webcams) in love, sex and romance? How important is physical touch? Other researchers (e.g. Radde-Antweiler 2007; Todd 2015) have discovered cohorts, for example those who play online games such as Second Life and World of Warcraft who appear comfortable with online relationships but I was not sure about my group of participants. I begin with Katie's story.

Katie's story

Katie was one of the few participants who spoke freely about her romantic and sexual relationship on Skype. The interviewer began by asking her if she had had an intimate encounter on Skype, perhaps an exchange of tenderness which might be sexual or a shared secret. Katie responds:

> Yeah, I think I've had all of those things via Skype; yeah. When I first moved over [from New Zealand to Australia] as I said, still close with Sam, my ex, and we were still sort of seeing each other. No we weren't 'cause we were in different countries but we sort of maintained our relationship via Skype. So yeah ... It helped at the time because it made us feel close and we can see the other person and that, and even though it was bitter sweet because we missed each other, it was nice to be able to see him smile or to see his quirky little mannerisms and call each other our pet names face to face, whereas over the phone you're a bit more distant. See it was nice but I guess it does kind of help maintain a relationship. So when I wanted to really end that relationship Skype was one of the first things that I stopped doing because it was prolonging it and it was giving him the wrong idea and he would see me and miss me more; whereas I was sort of seeing him and thinking, yeah, I really don't care about you but I wanna move on now. So I had to stop Skyping to break down that relationship into more of a friendship.

The interviewer then changed tack slightly to ask Katie whether, when she Skypes, she likes to just see people's heads and shoulders (the perspective commonly adopted still in many Skype interactions) or if she would prefer to see more of the person.

> KATIE: No, I'm quite happy just to see their head and shoulders. I don't really know that I'd wanna see more than that. It depends who it is again. If I saw someone who was just a friend [she pauses] if you're in an intimate relationship with someone or whatever, it's fine to see the rest of their body on Skype; but if it's just a friend I wouldn't be interested in seeing anything but their head and face, and in fact I'd probably feel a little bit strange seeing their whole body on Skype or even their whole waist up. I'd sort of think ... it would be too much. It would be too intense because it's not real life. In real life you expect to see someone and their whole body obviously but I think it would just be a bit too much even being in an intimate relationship with someone and maintain that relationship over Skype. When I was in a long distance relationship in New Zealand and I was in Hamilton and [name of boyfriend] was in Auckland, we would Skype and it was fine to see a part of his body like his head or his chest or something like that, but anywhere below that I was kinda like, 'Get it off the camera.' Do you know what I mean? Any relationship [she

pauses again] naked and he'd get naked on Skype and it would be a bit of fun for us and like, 'Oh hi,' 'cause we couldn't see each other all the time; we were only seeing each other once a week or once every two weeks. But it just kind of made me feel a bit uncomfortable because I just don't think Skype is the place to expose yourself.

INTERVIEWER: Well, everybody would have different preferences. I'm sure there's plenty of people out there that find it quirky and fun and have no problem with it whatsoever; but for some people intimate encounters are better in the flesh.

KATIE: Yeah. Yeah, that's it. It depends how comfortable you are.

INTERVIEWER: Was [name of boyfriend] more comfortable than you, would you say that, in doing that?

KATIE: Yeah, definitely. Yeah, he was. He'd always be up to be naughty on Skype and stuff and I'm kind of like, yeah, sometimes I'd go along with it and be like yeah, okay, whatever. I'm in Australia, who cares? But I would never put myself in a position where he could see my whole body – my face all the way down to my legs – it would sort of be in bits and pieces because I wasn't comfortable being that exposed on Skype with my whole body. And so when he would do those kinds of things I'd be like, oh, too much.

INTERVIEWER: That's really interesting. So for you is it about 'Big Brother' watching or is it just a privacy thing? When I say 'Big Brother' watching I mean there have been incidents in the news where government agencies have tapped into things like Skype.

KATIE: I didn't know that.

INTERVIEWER: Oh sorry. But I was just wondering what drove those privacy or inhibitions for you as opposed to [name of boyfriend] who was quite carefree with it?

KATIE: I think to me it just didn't feel natural being on a video and doing those things; but [name of boyfriend] is a lot more liberal than me in real life. When he was a uni[versity] student he went straight through the halls butt naked. He was always getting naked all the time. All of my friends had seen him naked and I don't care, that's fine for him, but I'm not gonna do that; it just doesn't feel, it's not the space to be naked in, completely naked, for me anyway. I don't wanna be that naked.

Miller and Sinanan (2014, 67) make the point that: 'The absence of sex is potentially a hugely damaging and debilitating consequence of ... [a couple's] separation, which has become an increasingly common consequence of global migration'. Skype provides something of a solution but as can be seen from Katie's comments, it is not a solution that satisfies everyone. She did not want to expose all of herself, nor did she want her partner to expose all of himself – 'bits and pieces' were okay but the 'whole body' was 'too much'. In the future people will likely continue to be mobile and more partners will live apart. The issue, therefore, of how Skype might be able to help deserves more attention.

If Skype is able to assist in bringing people emotionally closer together, just like it is bringing grandparents and children emotionally closer together, then this could potentially be a positive development. Despite this logic participants in the research still far preferred co-present sexual relations. Anything else involving a webcam for most inferred sexual relations that are 'lesser than' in terms of satisfaction, morality and authenticity than bodies actually physically touching. In considering Katie's comments through the lens of gender relations, Nicola Döring (2000) argues that feminists tend to read 'cybersex' through two contrasting lenses, one that sees it as potentially liberating for women, freeing them to explore their sexual identities and the other as a heterosexist practice that is harmful to women (also see Gibson 2004 on 'gender, pornography and power'). Katie was less comfortable than her boyfriend being naked in front of the webcam. While she felt comfortable showing her boyfriend particular sections of her body at different times, she felt 'exposed' showing him her whole body. While Katie did not explain why exactly, similar discussions on Reddit indicate that many women tend to feel it is easier to frame particular bits of their body as 'sexy' whereas the whole body framed can appear very 'real' and even somewhat confronting in that for most it does not look at all like the highly reproduced, airbrushed photos of women so many are accustomed to in the glossy magazines.

Real sex and contrived sex

While Katie was open to the possibility of Skype sex most were not. Some felt more strongly than others that somehow Skype sex wasn't real.[1] For example, Barbara, a medical receptionist, was totally opposed. When asked if she had been intimately involved with her husband via Skype, she replies:

> No way, nah; that sort of thing freaks me out. Nah, I'm not into that ... there's no way I would do that, you know how people do 'sexting' and all that sort of shit – no way man! Even though Skype's probably more private because it's just between you and him, or the other person, I still wouldn't do it. I would not feel comfortable doing that at all!

Barbara, after explaining this, called out to her husband in the next room saying: 'She [the researcher] wants to know if we would have Skype sex?' Barbara's husband, a miner, retorts: 'Skype sex? Jesus, I can't even get real sex let alone get that!' [laughter]. Barbara adds: 'No, nah, maybe if we were younger, but nah; that's just [voice trails off]. No I wouldn't.'

Butler (1993, xi) argues that there is no doubt that 'bodies live and die; eat and sleep; feel pain, pleasure; endure illness and violence' and that these cannot be 'dismissed as mere construction' (ibid.) but she asks 'why is it that what is constructed is understood as an artificial and dispensable character?' (ibid.). Barbara's husband refers to 'real sex', meaning sex that involves him and his wife being physically intimate. He implies that sex involving images on a screen is (unlike 'real sex') constructed and therefore artificial and dispensable. He also implies that it is a little 'on the wild side' – an activity that is beyond the bounds of their marriage.

When I asked Debbie, a university lecturer, about whether she had engaged in sexual relations with her partner via Skype she was not as opposed to the notion as Barbara but did not really see it as a possibility.

> No, no, we haven't done that ... 'cause I'm quite a huggy person. I'm always very open to that kind of physical part of a friendship; embracing or whatever ... It's an interesting question ... it's maybe part of why I find it [Skype] a bit distancing and if it was me having a sexual relationship through Skype I probably wouldn't choose Skype to do that. I don't think that's my medium ... it's a poor substitute.

Skype for Debbie is a 'poor substitute' for what she considers to be real (read: offline) intimacy while Skype enables a range of other bodily senses to be engaged such as the visual and aural it does not enable touch and smell (see Gilbert, Murphy and Avalos 2011 on 'three-dimensional versus real-life intimate relationships'). This can tend to result in the visual being highlighted to the point where a number of participants felt it became voyeuristic, too much about looking and not enough about the other senses.

Ahmed (2006, 66) argues that queer moments in which objects appear 'out of line' function to 'block bodily action'. Skype blocks some bodily actions. For example, when one sees the body framed differently and/or the visual appears to dominate, and when one attempts to touch the body of another but that body turns out to be beyond physical reach then sexual desire may be queered or blocked. Skype presents a different way of being sexually. Ahmed (2006, 67) points out that 'when Merleau-Ponty discusses queer effects he is not considering "queer" as a sexual orientation'. Similarly, when Ahmed discusses what it means to 'orientate' oneself sexually towards some but not others she is likely not thinking about Skype sex and the orientation of images through cables, computers and screens but nevertheless her research offers some insights which can help make sense of participant comments.

When I asked Beverly, aged in her late twenties, 'Have you ever had any intimate encounters via Skype – romance, tenderness, shared secrets, intimacy, that kind of thing?' she indicates that she would not have sex via Skype because it is not the kind of sex that she wants – it's 'voyeuristic and exhibitionist' she says:

> When I Skype with my partner we would talk about things ... that involve a certain degree of intimacy. If we're talking video sex, no! ... I've never done it. I can see how it could be useful in long term, long distance relationships but ... it's a particular kind of sex, that is quite a lot more voyeuristic and exhibitionist than anything I would ordinarily do. So it would involve thinking about my sex life in quite a different way.

This final comment from Beverly about thinking about her sex life in a different way and her use of the term 'ordinarily' indicate that Skype sex for her is not completely outside the realm of possibility but currently it is not something she feels at ease with it due to what she perceives to be it voyeuristic and exhibitionist nature.

Ewan, a government employee aged in his early fifties, concurs with Beverly. He says:

> It seems so contrived, like play acting somehow. It seems to me, to be, not what our relationship is about. I know that sex is [pause] there's performance in sex, but it somehow, there's a lack of [pause] I was going to say dignity, but there's nothing particularly dignified about sex! Integrity I suppose, somehow.

When I ask Ewan to explain further he adds:

> It's just not feeling comfortable; I just don't feel comfortable enough. I feel tense around the camera, I feel awkward about myself, so that's not going to go well for a sexual encounter.

Debbie, Beverly and Ewan all share in common a feeling that while they are not opposed to others having sex using Skype – they do not necessarily have a moral objection to it – it is more that for them they do not like the way Skype highlights the visual over touch. To return to the opening story about Katie, she too found sex using Skype to be uncomfortable. While she enjoyed the tenderness explaining: 'it was nice to be able to see him smile or to see his quirky little mannerisms and call each other our pet names face to face', she was not comfortable seeing all of her boyfriend's body on the screen or him seeing hers, saying 'I just don't think Skype is the place to expose yourself' and that to her 'it just didn't feel natural being on a video and doing those things'. They all appeared to feel that Skype sex involves a sort of staged bodily performance that has links to pornography and masturbation with which they feel uncomfortable. Miller and Sinanan (2014, 63) sum up: 'Webcam simply doesn't work for everyone, and some don't find it a comfortable medium for intimacy'.

Brittney, aged in her early twenties, had used Skype for romance with her boyfriend but not for sex, or at least not that she spoke about in the interviews. Brittney lives in Hamilton, Aotearoa New Zealand. Her boyfriend lives in Auckland, one and a half hours drive to the north. She lives with a couple of flatmates and has been using Skype for just six months. Unlike most of the other participants in this study Brittney live-streams Skype with her boyfriend for long periods. Miller and Sinanan (2014) suggest that this increasingly is the way people are using Skype as they become more accustomed to it. The technology is beginning to blend into the background.

BRITTNEY: Sometimes me and my boyfriend will just have Skype on but we'll be doing other things and then we just talk and then we can see each other you know and the little box will be up there if we're doing something else. And actually when I think about it we did find some things to do to try and keep our relationship, so we synced a movie together and we watched that together. That was pretty difficult because we had to press play at the same time.

INTERVIEWER: Was that on the computer or on like a TV that you were doing that?

BRITTNEY: No, so on the computer. We got the movie that we wanted to watch and we put our boxes up in the corner and then we went one, two, three, play and we had to sync; so each of us had the movie going on our computers and that was interesting and that was quite cool.

INTERVIEWER: Did that did kind of work?

BRITTNEY: Yeah. I think in the end it got out of sync and he just muted his and just listened to my one I think; I think that works. And then we had like a cooking date where I just had my computer up while we both cooked something and we just talked about what we were cooking and then we ate together and stuff like that.

INTERVIEWER: In terms of the movie experience how is that different to actually being in the same room watching a movie together?

BRITTNEY: I don't know, I didn't actually find it too much difference apart from actually feeling the other person with you; and it's kind of exciting.

INTERVIEWER: Exciting because it was something different?

BRITTNEY: Different yeah.

INTERVIEWER: It wasn't just talking about random stuff?

BRITTNEY: Yeah.

INTERVIEWER: It was actually doing.

BRITTNEY: And then like when things happen we can still talk to each other.

The interview with Brittney indicates that Donna Haraway's (1990, 220) question 'Why should our bodies end at the skin?' seems more relevant today than ever as technologies such as Skype, especially when used to live-stream for long periods, illustrate that bodies, boundaries and spaces are increasingly merging and challenging normative performances of how people are used to being.

While Brittney has succeeded in finding new ways of being in relation to spending time with her boyfriend in another city, for example, watching a movie together using Skype, most of the participants in this study did not live-steam absent partners. Nor did they feel that online romance and sex were as real as offline. This finding, however, is not routinely echoed in the research of others. Some studies (e.g. Ben-Ze'ev 2004; Todd 2015; Underwood and Findlay 2004) have shown that there are people who report that their online experiences of love and romance feel just as real, and in some cases, more real than offline romantic relationships. Aaron Ben-Ze'ev (2004, 185) argues that people involved in online relationships 'are often highly emotional and sexual'. Cherie Todd (2015) in her study relationships between gamers in the online game World of Warcraft found that many of her participants had established deep emotional and sexual relationships with other gamers, some of whom had never met offline. A study of the online game Second Life (SL) by Richard Kolotkin et al. (2012, 1) found that many of their participants had very satisfying online relationships. This was not the case for my participants, however, who for most part viewed emotional and physical relationships conducted via Skype as being somehow less real, more contrived than offline relationships. Seeing and hearing but not being able to touch one's partner left them feeling as though intimacy was somehow lacking.

Often before technologies are embedded as a routine part of individuals' everyday lives there is a sense of discomfort with the medium. Miller and Sinanan (2014, 3) argue that 'there are no unmediated, pure relationships. All the ways in which relationships exist including communication, are cultural activities.' This makes sense and yet most of the participants repeatedly indicated that they felt Skype mediated intimacy, especially sexual intimacy in ways that make it feel different. As time goes on, however, and people become increasingly familiar with synchronous audio-visual mediums such as Skype, perhaps leaving it turned on for long periods during the day and night, what is considered pure, authentic and unmediated will likely shift.

What constitutes a real relationship – i.e. being face to face, or perhaps more appropriately for this chapter, being body to body – is likely to be increasingly questioned in the future. Quentin explains that while working away from home he was keen to engage in Skype sex but his partner has a 'disdain for screen technology' and so it has not happened. He says:

> It's not through a lack of wanting, it's just that my partner is not into technology, right across the board, and not into screens, and tries to avoid screens where possible. So I think that's the only barrier to that happening. It's not like there's any lack of will on her part.

Quentin explains that on Skype he has had 'tender moments, definitely shared secrets' to the extent that he has had to delete his instant messaging history in case a third party found out what had been discussed but not sexual intimacy. He explains 'definitely the will is there but I've never had that'. Instead 'sexual intimacy has occurred via the telephone' because as mentioned earlier his partner is not keen on facing a computer screen. He adds: 'Yeah it's kind of something I'm totally into but it hasn't … as we discussed earlier, just hasn't transpired.' Again, there is evidence here that not everyone is comfortable with the visual aspect of Skype for sexual intimacy. Quentin is comfortable extending his desire, his body into the virtual space of Skype in order to get near to the body of his partner. She, however, is not prepared to 'deviate and pervert the lines of desire' (Ahmed 2006, 105) which she feels will happen if she faces the computer screen for sex. The screen is a barrier for Quentin's partner, a barrier that she is not willing to cross.

Generational difference?

In this research, gender appeared as an area of difference in relation to perceptions of Skype sex with men appearing more willing than women to engage but so too did age or generation. Barbara, aged in her early fifties, associated Skype sex with the younger generation. As mentioned earlier, Barbara says 'maybe if we were younger'.

Kurt also associates Skype sex with a younger generation. He explains that he would not consider having 'cybersex' (as he described it) with his wife, adding 'I'm too old for that [laughs].' Bryan, aged in his early seventies, when asked about the issue replies:

> Yes, people who want to communicate with each other in more intimate ways who might do something that most of us would find questionable ... For me I just don't get the current practice of sending [risqué] selfies to other people ... People do extraordinary things with their control of technology which I find just appalling.

Although Bryan did not state explicitly, I got the sense from the interview that he was meaning *young* people do extraordinary things with their control of technology.

At another interview, Ewan aged in his mid-fifties, I had earlier in the day seen an advertisement for the 'Svakom Gaga – the world's first camera vibrator'. The conversation seemed rather relaxed so I mentioned this. By way of background information, the advertisement for the vibrator reads: 'Ladies – ever wondered what your insides look like when you orgasm? Well wonder no more.' The product is fitted with an inbuilt high definition camera. Dubbed the 'X-rated selfie stick' it promises 'to take intimate snaps and home movies to the next level' by women being able upload images to their PC or smart phone, or synch with FaceTime (May 2014). The advertisers enthuse: 'Like all good selfies, you don't need to keep these images to yourself. Thanks to its FaceTime compatibility, sharing the view with your lover couldn't be easier – whether you're sat right next to each other or miles apart' (Goorwich 2015). When I explained this product to Ewan he responded that while it might make sense to the younger 'selfie generation' to him it sounded more like a medical procedure than sex.

The mean age of participants in this study was 40–44. It is tempting to think, like Ewan, that a generation gap of sorts might be in play here. I wondered if interviewing a younger cohort of participants might have produced different results. Would younger people have been more comfortable with using Skype and other audio-visual technologies as part of their sex life? It is possible but certainly not a given. A group of three friends, Santino, Daina and Alex, all aged 20–24, who were interviewed together said, like many of the older participants, they often found Skyping 'awkward' (and not just for sex, love and romance). Alex explains that one of their other friends would 'like have a party with friends in other cities ... have like drunk Skype parties' to which Daina responds: 'Lots of people do that.' Alex adds: 'I've been drunk on Skype ... I think it made it way less awkward than usual. I think that was the best time I've ever Skyped; it was like an hour long.' Despite the rhetoric in popular culture that young people are all 'digital natives' and entirely comfortable with all digital technologies including webcams this is not necessarily the case.

In seeking to establish a broader context in which to place the interview data I reviewed at least 30 conversational threads revealed when searching 'Skype sex' on Reddit. In the first instance the entries turned up by this search were broad-ranging but I focused on sexual and romantic relationships between partners, lovers, husbands and wives rather than strangers 'hooking up'. In these threads there are many people seeking advice about 'how to make it [Skype sex] work'. Oftentimes it appears that one partner (frequently, but certainly not always men) is keen but the other is not. There are many stories of discomfort, awkwardness and seeming failure. Many women still appear to associate webcam with pornography and feel that being involved may somehow sully them. One writer explains:

> We've been long distance for about 5 months now and after spending a whole month together we thought we would try and deal with our sexual frustrations the most intimate way we could think of given the distance: skype sex. She ordered a vibrator and it arrived today, and we tried... I eased her into it with some dirty talk and tried to get her as comfortable as possible but after 20 minutes of not being able to cum she broke down in tears, told me she loved me and rushed off to the shower.

Another contributor asks readers:

> I have sort of tip toed around skype sex as a sort of laughable, yet maybe somewhat plausible thing we might want to try but are still very hesitant about [it] for a while now. We've been together 2 and a half years and are very comfortably and happily sexually active when were together, but honestly, speaking only for me, I don't see how skype sex (mutual mastrubation [sic], whatever you wanna call it) can be anything but awkward. Am I just not getting it?

Others have more positive experiences to share. Similar themes under the headings relationships and LongDistance emerge to those that emerged when talking to my 39 participants, that is, feeling it is Skype sex is somehow lesser than real sex because of the lack of physical touch and an awkwardness about performing on camera.

Another issue that emerged both on Reddit and amongst participants was whether one ought to trust a partner not to record intimate Skype sessions including screen shots and video capture are also raised on Reddit. So too are issues of someone 'hacking' into the computer over the network to watch. One of my interviewees also raised this. Kathryn, a psychologist and her partner Harrold, a boilermaker and welder, spend time apart when he is onsite at a rig. They were interviewed together and explained that they have joked about having sex via Skype but not done it. Harrold says:

> Well of course they [Microsoft] won't admit it but I think that any business that is involved in social media is mining their data ... the MI6 or MI5 or something, they got caught out for recording. I think it could have been screen captures or images off Skype and 10 or 20 per cent of them were of a sexual nature, contained 'inappropriate nudity', I think was the term they used. And so I'm aware that anything that's been said over Skype could then be monitored at any stage.

Skype originally featured a hybrid peer-to-peer and client–server system, however, since May 2012, it has been entirely powered by Microsoft-operated supernodes. The 2013 mass surveillance disclosures revealed that Microsoft has granted intelligence agencies unfettered access to supernodes and Skype communication content (Wikipedia 2015a, Skype.com 2015). The increasing potential for surveillance culture is an issue that has emerged as Skype as grown in popularity and is maybe one of but not the only reason that participants in this study regardless of age, were not very comfortable 'sinking' into the romantic or sexual encounters using Skype. Security issues aside, it seems that most see these encounters, that is, these 'queer moments' (Ahmed 2006, 65), as unnatural. Interestingly, the only participant for whom this was not the case was Imogen, a performance artist and lesbian, who has been using audio-visual platforms for a very long time – even before Skype existed. Izzie says:

> Before Skype I did have one kind of sort of romance I suppose with someone in New York and that was using I-Visit [real-time video conferencing over the internet]. We'd met through websites and emails and then I can't remember exactly at what point we decided to start talking on 'I-visit' but I was using it at that time for performances so I had it and I told her how to do it and how to set it up. It was a short period but we had some quite interesting intense – they weren't Skype chats – but conferences. There was one other person that I was a bit romantically involved with that we had a little brief thing also – I wonder whether that was even with Skype; no it won't have been Skype it was too early, it was 2003. We must have used I-Visit as well ... I think for me actually a lot of it is to do with words and being kind of poetic images. And you know like I have in those situations set something up but something that I guess was probably more kind of artistic. I remember one time I did a thing; I had a pair of long black velvet gloves and I did things around the edge of the frame and I did things just with my hands and taking the gloves off and on ... I lit it nicely and I had a bit of fabric at the back or something,

so I made a kind of little stage and I made a little performance. It was quite intimate and it was just for that person. I just had the idea and I didn't rehearse it or anything; I had the idea before we talked and then I said, 'I want to try this,' and I did it for them. But I don't know that everybody would do that kind of thing or necessarily find it [voice trails off]. But I think if you need to go and look at a website to get tips on that kind of thing then you haven't got very much imagination. I mean there are a lot of things you can do with a camera [laughter].

Over the past decade, the spread of digital media has profoundly reshaped intimate relations worldwide, transforming the ways in which many, but not all, people experience sex, love and romance in their everyday lives. Alan McKee, Rian McNair and Anne-Frances Watson (2015, 159) note 'the online environment [for sex] ... may be navigated with the aid of no more than a computer or smart phone and an Internet connection'. In talking about sex, love and romance via Skype I am not implying that digital media have inherent capacities or qualities that necessarily determine different emotional and affectual outcomes (Bingham,1996, 2005) but they play at least some role (Baym 2010; Benski and Fisher 2013; Longhurst 2015; Clough 2000). Whether media are voice-based, text-based, multi-media, synchronous, or asynchronous can make a difference to the interactions just as real space makes a difference (but again does not necessarily determine) interactions.

At the outset of this chapter I asked how successful is Skype with video in providing an environment for expressing a variety of different kinds of love, sex and romance? Participants reported that while Skype enabled them to share intimate, loving or close moments which are aided by the webcam when it comes to sex the webcam tends instead to prompt feelings of discomfort and awkwardness. The majority of the participants in this study still felt highly self-conscious in front of a webcam when it came to sexual performance. So, rather than *seeing* the other person aiding romance and sex it tended to hinder it. While others, particularly those focusing on online gaming have uncovered other results, my finding is supported instead by Miller and Sinanan (2014) who report in their study of webcam use in Trinidad that only a few of their participants (couples) were attempting to use visual technology for romance such as online 'dates' (e.g. simultaneous watching a movie or television series). Their overall impression was that dating, and I would add here, sex, via webcam 'remained somewhat awkward and not overly successful' (Miller and Sinanan 2014, 56).

Another question posed was how important is physical touch? For participants in this study touch was very important. At one level it makes little sense to extract touch out from other bodily senses and yet touch – skin-to-skin, friction, the tactile, merging of bodies – for most was considered to be integral to sexual relations. Smell too is a sense that is missing in Skype encounters although this was not mentioned by participants in relation to sex. Emotion and affect transfer are different via Skype. The capacity of bodies to impact on each other through a screen rather than in offline space prompts the enactment of different multi-sensual selves. It is possible to see and hear people but not to touch and smell them. In the process selves become reconfigured. Sex is still sex, but not as many know it.

In this way Skype is a space that has the capacity to queer social relations. By this I mean, at this particular historical juncture, it is un-fixing how many understand intimacy. It is making the familiar strange. Sexual acts have long been disciplined by social institutions and practices that normalize and naturalize certain types of sex. Sex via a camera and screen carries with it for many connotations of being 'unnatural', beyond the norm, something Others do, for example people who pay for sex, have a voracious sexual appetite, or are immoral in some way. Miller and Sinanan (2014, 56) explain: 'At the moment, we may be in a more experimental period' of webcam use. When new media first begin to be used we tend to do so as the more recent iteration of some prior media. For example, webcams tend to be used more as telephones. We phone people to talk for a set period, only now we have an image as well as voice.

Some people, however, are beginning to use webcams and media such as Skype differently, keeping them on for long periods while simply going about their everyday lives. This may lead in the future to people feeling more comfortable with having loving, romantic and sexual interactions online, via webcam. Currently we may be in a space of in-betweenness, a queer moment in which people are reasonably familiar with the technology but not overly so. 'Netiquettes' are yet to be established between partners (Helsper and Whitty 2010). They are not quite sure what constitutes intimacy and indeed what constitutes 'internet infidelity' (Hertlein and Sendak 2007).

The technological and cultural are merging in interesting ways that may enable us to enact change, that is, to prompt different sexed and gendered ways of being that move away from current norms. I am not implying that synchronous audio-visual technologies will lead to some kind of utopian sexed and gendered future but as Michele White (2015) argues femininity (and arguably masculinity too) are being shaped in new ways through new media. There are possibilities for reiterating current sexed and gendered practices but maybe there are also possibilities for new ways of doing sex and gender in ways that queer the subject. No matter whether sexual desire orientates the subject towards the 'same sex' or 'other sex' (or any number of possible orientations that well exceed any two-sexed model) the issue here, at least in part, is the matter of the screen. Many align arousal with touch and to deviate from this (for it to become about representation and image) is queer, pornographic, immoral and masturbatory. Online is seen to be 'off line', that is, to not follow the line or the usual way of doing sex (Ahmed 2006).

Ahmed (2006, 78) explains 'perversion', like 'queer', is a spatial term which can refer to 'the willful determination to counter or go against orthodoxy, but also to what is wayward and thus "*turned away* from what is right, good, and proper"' (italics in original). For many of the participants being sexually intimate through Skype seemed somehow perverse, to fly in the face of social convention turning the computer, laptop or mobile device being used into an object that could prompt disgust.

Another way in which Skype sex is constructed as queer is that unlike real or 'natural' sex it does not involve penile penetration of a woman's body, it will more likely involve self-touching and perhaps sexual objects to pleasure the body. Sedgwick (1993, 112) notes that there is a 'history of masturbation phobia' constructing it as, at worst degenerate, and at best pathetic. It is also constructed as a little queer and 'unnatural' – not resulting in procreation – 'the gift of life itself' – and all that that entails by way of paternal lines of descent and inheritance (Ahmed 2006, 86). Over time as people begin to increasingly use Skype for sexual encounters, repetitive performances are likely to gather force to produce the appearance that Skype sex is 'natural', but for now at least, to many it still seems like an 'unnatural' orientation of bodies in space.

Note

1 See Parks and Roberts (1998) on developing online and offline relationships. For more on the 'realness' of sex in relation to cybercheating see Whitty (2003, 2005); Whitty and Carr (2005 , 2006) and Whitty and Joinson (2009).

8 Reorientating bodies and spaces

And so, what is one to make of the various stories from participants, information from a range of internet sources and Ahmed's queer phenomenology? The task in this final chapter is to bring together these various lines of enquiry and make some sense of them. James Ash in The Interface Envelope (2015b) draws on the example of videogames to make the point that interfaces create envelopes, or localized foldings of space and time. These foldings of space and time, Ash argues, create environments that organize players' perceptions in such a way that the games succeed at generating economic profit – a sort of 'cognitive capitalism'. Skype is both similar and different in this regard. The platform, particularly when used for commercial purposes (Skype for Business) undoubtedly is marketed with the aim of generating profit. As noted in Chapter 1 Skype was acquired by Microsoft in 2011 for $8.5 billion. Individuals for personal use, however, tend to spend little or no money on Skype. The software is available to download for free and there is no cost for Skype-to-Skype calls. All users need to purchase a computer or some other device and pay fees to telecommunication companies for internet services in order to use Skype but for personal Skype to Skype, calls with or without video are free.

This means that while some parallels can be drawn between this research and Ash's (2015b) study of the worlds created by videogame designers where the explicit aim is to extract profit from users, it is something of a different 'folding of time and space' that occurs. There is no doubt that Skype is in the business of selling, ensuring their service is an integral part of people's personal and work lives but there is something more I have tried to get at in this book about its use in different contexts for different purposes. Ash's (2015b) interfaces which create envelopes of power are useful for understanding the computer game industry and how they organize players' perceptions but I want to return here to lines of sight/site – straight ones and queer ones – for understanding people's experiences of Skyping friends, family, loved ones, employers, colleagues and lovers.

Lines of sight/site

By using the term sight/site here I want to highlight that in thinking about lines of sight I am referring to visual lines – what can be seen via the webcam on the screen in the foreground and background – but also lines of site – power, cables, cords – which connect online and offline spaces. Skype is creating new lines of sight/site which are orientating but also disorientating bodies in particular directions, shaping what bodies can and cannot do and who they can and cannot connect with. Ahmed (2006, 157) argues that: 'Moments of disorientation are vital. They are bodily experiences that throw the world up, or throw the body from its ground.' And, as Ahmed continues, these feelings of disorientation can be unsettling. Sometimes the work of inhabiting new spaces is done willingly, for example, when we desire to use Skype to connect with friends who we miss because they live a long way away even when we know the technology might fail. Other times we are forced to engage with new spaces, for example, when our employer has decided they now want to routinely Skype us with video but we would rather not be receiving these calls.

When bodies create new lines it can result in feelings of disorientation. It can, however, also result in feelings of being 'at home', comfort, or 'sinking', especially after some time has passed and the bodies have grown accustomed to the new lines and new spaces. For example, grandparents enjoying listening and watching grandchildren show them pictures or other objects on the screen.

This of course is not a binary – disorientated *or* orientated, uncomfortable *or* comfortable. There are multiple bodies, objects and spaces. Becoming comfortable in a greater range of spaces, such as Skyping at funerals, for a job interview, or for sexual gratification will take more individual and collective work. As that work takes place bodies will acquire new shapes and the spaces and places in which they dwell will continue to change.

New power relations will also keep folding into these newly created spaces. For example, paid employment will likely increasingly permeate home. Employers will likely increasingly want employees to 'stay in touch' not just by texting or voice-calling but also by video. This risks workers' rights being infringed as they feel increasingly under surveillance through Skype (Meikle 2016) and other synchronous audio-visual platforms in their workplace, outside their workplace and in their homes. Doreen Massey (1996) in relation to workers in the high-tech sector in Cambridge, England explains how work is taken into the private spaces of the home as employees work long hours but it is less acceptable for home relations to cross the boundary into spaces of paid employment. Changes in technology can lead to changes in power relations and orientations and these are often classed, raced and gendered.

Through seemingly ordinary moments in everyday life as people Skype each other at home and at work, in private and in public, spaces are acquiring a new shape. Each of the chapters in this book focuses on different aspects of subjectivity such as being a grandparent, family member, friend, paid worker, partner or lover – to illustrate different lines of sight/site between bodies, emotion, affect, materiality, objects, power, space and time. The aim was not to simply offer up theory for theory's sake but to see what Ahmed's queer phenomenology might be able to 'bring to the table'. I wondered what might happen if I brought it into conversation with the experiences of participants in the project.

What has transpired is that Skype is not a new digital media that has brought about a utopian future for all. It is often lauded by its marketers, and by some users who reinforce this dominant discourse, for its capacity to connect people for the purposes of both pleasure and business thereby addressing some of the shortcomings of real space and time. Skype is popularly conceived as a marker of progress towards a better, closer and more integrated world. People across the globe connect with each other, interacting in seamless fashion made possible through advances in technology. And yet as most of us are only too aware, it does not provide a perfect solution for people connecting. The body seen on the screen is familiar in its look and comportment but we cannot touch it, feel it, in the way we might be used to. There are likely advantages and disadvantages to this depending on the context but what is clear is that it is a little queer.

The lines of sight/site are not clear. While Skype has not brought about a utopian future, neither has it charted the terrible path described by those who think digital media are 'taking over' from real or face-to-face interactions which are always much more authentic and meaningful. Face-to-face interactions may enable people to see, touch and feel in ways that are familiar but they are still mediated. Lines of sight/site still have specific projections, foregrounds and backgrounds that must be negotiated as part of the interaction.

Constructing a binary of face-to-face as a real or true form of sociality and Skype as a more mediated and therefore lesser form of sociality makes little sense given that all communication is mediated in some way. As Ahmed (2006) reminds us, it is about how bodies are situated or orientated in space that matters not degrees of mediation.

Queer phenomenology then opens up ways of going beyond these reflections of degrees of mediation to instead think more about some of the complex social aspects of relationships between people and spaces. Using Skype involves feelings of both pleasure and unease, familiarity and unfamiliarity about both the self and others on the screen. Sometimes bodies on Skype feel like they are not quite in the right place, other times they feel perfectly positioned. Sometimes they register as a sensory intrusion, other times as a sensory lack. This can be disorientating – it is not what we expect. Disorientating moments, says Ahmed (2006) are moments where we lose perspective, we are dislodged from the usual dimension we inhabit and come to inhabit a new dimension. Skype is a new dimension and there are things in this new dimension that we are beginning to grow accustomed to such as seeing our own image in the small box on the side.

As discussed in Chapter 4, the small box that reflects our image back to ourselves can feel like a mirror in which we want to fix our hair or check there is nothing between our teeth. We often do not know whether to look at ourselves, the images of others on the screen or the camera. Latour (1993, 51) notes that objects and subjects are always 'quasi' objects and 'quasi' subjects which form 'strange new hybrids'. The entangling of physical bodies with technologies to produce reflections of ourselves and others enable us to feel at times orientated but at other times disoriented.

Skype with video means that people are become increasingly visible and therefore identifiable in newly defined virtual public spaces. It is becoming increasingly normative for people to turn on their webcam thereby surveying themselves and others. There are, however, some who wish to hold back, at least parts of themselves, from this mandate of visibility. There are times when people do not want to be seen because it feels intrusive. It is likely that in the future this may increasingly come to be seen as suspicious and Skype callers will begin to wonder 'why don't you want to turn your video on?' Miller and Sinanan (2014, 167) explain that there is now a

> presumption that if a truth is now embodied in the very existence and possibility of webcam, then it becomes wrong, a falsehood, a sin, to be talking on Skype without the webcam actually on ... To refuse webcam and neglect that capacity to be seen can become a judgement upon you, imputing the motives behind this preference.

Miller and Sinanan make a useful point and one that can be reiterated by directing attention back to queer theory, but this time to the concept of 'the closet' (which Ahmed (2006) notes is itself an orientation device). Increasingly there is an expectation that bodies are being expected to come out of a closet of sorts, that is, for their bodily materiality to be revealed on screen. Brown (2000) argues: 'the closet is not just a metaphor for the concealment, erasure, or ignorance of gays' sexualities.'[1] The closet exists at a variety of spatial scales, from the body to the globe. The closet can also exist offline and online.

While conducting this research one of the participants, a woman in her late fifties, agreed to do a Skype interview but when the arranged day and time arrived she did not want to turn her video on even though I had explained I was interested in the visual aspects of Skype. I had a strong sense that she preferred to stay closeted. The participant could see me and commented that I looked 'lovely', 'nice', 'cool', 'an attractive lady' and that she 'loved' my background because it was 'arty and very professional looking' and 'matched my face and hair' but said she did not know how I would feel if I could see her. People coming in too closely into her world made her feel 'nervous' and 'very uncomfortable'. She told me that she turned her video on to Skype the women she employs but with others she prefers an audio only call. Feeling like one does not look attractive enough, is insufficiently groomed, does not have a professional or enticing background might mean that they do not want to reveal themselves and where they are. This participant felt embarrassed, shame even (Probyn 2005), preferring to remain closeted.[2]

There are, however, mounting pressures for people to develop and maintain linkages via Skype, and not just by way of audio but also video calls. Graham Meikle (2016, 123) describes this as 'distributed citizenship', that is, the taking up of 'a creative political relation with one's contemporaries within social media networks.' While this concept might perhaps be more obvious in relation to social media such as Facebook (e.g. those who are not on Facebook may increasingly feel out of the loop of political and social media flows) the point can also be made in relation to Skype. To gain 'cultural citizenship' one must have a Skype account, upload a photograph in one's profile, and keep adding to one's contact list (using functions such as 'Find Facebook friends'). Being 'out' then involves not just revealing oneself visually to those we call but also as Graham Meikle (2016, 124) argues, it involves being in 'the domain of communicative environments which are largely under the control of enormous US media and technology corporations'.

While I have not in this book emphasized aspects of surveillance and commercial exploitation, these are nevertheless important aspects that Meikle (2016) and others pay attention to. The relations between state institutions, corporations and citizens are being fundamentally reconfigured and issues around identity, security, privacy and publicness are important. They are, however, not what was addressed or emerged in any great detail in the interviews I conducted. If this project had focused on a different social media such as Facebook, Twitter, or Instagram themes such Big Data, data archives, ethics, activism, digital literacy, digital divide, archiving and cyberbullying online might have emerged more clearly with the exchange and capture of more asynchronous text-based information.

I saw my task as a feminist geographer who is interested in the body as being able to think through, with the help of Ahmed's queer phenomenology, whether Skype is prompting moments of disorientation and if so what might we do with these moments. Can they offer hope for new directions or lines of sight/site that lead to more a more socially just world? This question is not easy to answer because as Ahmed (2006, 159) argues:

> disorientation is unevenly distributed: some bodies more than others have their involvement in the world called into crisis. This shows us how the world itself is more 'involved' in some bodies than in others, as it takes such bodies as the contours of ordinary experience. It is not just that bodies are directed in specific ways, but that the world is shaped by the direction taken by some bodies more than others.

This point reminds us that not all Skype users are the same. Older bodies may have more difficulty learning the shape of this new world but age is just one aspect of subjectivity. The gender of bodies also matters in the way they are framed and read on screen. Socio-economic differences shape people's access to technology. All of these factors contribute towards the uneven distribution of use of Skype and feelings of disorientation.

But can feeling disorientated or like the space has been queered led to positive outcomes? Jackie Wykes (2014, 5) argues that: 'queering can work to destabilise normative categories and denaturalise dominant ways of seeing, doing and being'. Skype arguably in some (but not all) ways has denaturalized dominant ways of seeing, doing and being. People are learning to be with others through a digital interface. Intimacy might no longer involve touching another person but kissing a screen, or wearing a garment that touches the self remotely controlled by another. It might involve doing a job interview with no trousers or child-minding through a screen but phoning someone if a child has material needs that must be met. It might mean attending a wedding or funeral across the other side of the world at a different time of day/night from the other attendees 'on the ground'.

Queer phenomenology suggests that through Skype people and objects are no longer in their usual place – what is distant and what is proximate. Long-held binaries between private and public space are being further destabilized. One day when I Skyped my sister in Aotearoa New Zealand from the United Kingdom (it was night for her and day for me) from a busy hotel lobby in London she answered with video appearing in her nightie sitting up in bed. The stranger standing beside me was more interested in my laptop screen than his own. It is evident to many of us that time and space are being reconfigured via screens as people use Skype, Facetime and other synchronous audio-visual platforms in hotel lobbies, on trains, in cafes, on streets, in bed. Having visual, simultaneous interactions can prompt feelings of having to get dressed, do one's hair, select an 'appropriate' video chat space, check back-lighting, and be careful about webcam angles so as not to distort the face. As has been illustrated, there are highly sexed and gendered dimensions to these embodied performances and so in some ways it is perhaps easy to argue that Skype seems simply to be reinforcing existing power relations – online and offline spaces become mutually constituted with one reinforcing the other – but there may also be potential in this space, at this time, to do things differently.

This book has attempted to capture something of people's use of Skype now – 2016. This is significant because the dynamics of any media are different at its beginning than when it is a more mature technology. For example, the original television programming of the 1950s to 1970s and the way in which individuals and families watched is different than the commercial-entertainment paradigm that emerged in later years. Families began to watch less together and instead it came to provide for many a backdrop for life. With Skype it seems that there is a similar process going on. It is not yet a mature technology. Some people still seem highly conscious about the way they use it, especially women who often judge harshly their own appearance in the side-box. There is, however, quite a high degree of comfort with Skyping friends, family and loved ones but participants in this study were still not entirely comfortable using Skype with video for work, especially not for job interviews. They were even less comfortable using it for romance or for sexual purposes.

Skype is still new enough that participants did not think it was in any way off or unusual that I was conducting this research. They described their interactions to me in-depth. Some were still in awe of being able to actually see and talk with others in synchronous interactions across time and sometimes vast spaces. Some were also still struggling with the technological challenges presented by the medium at times. Technology improvements such as increased bandwidth mean that everyday usage has increased especially for point-to-point calls between two units and nearly all laptops, tablets and smart phones now have inbuilt webcams and microphones but there are still sometimes challenges of images being pixelated or frozen, not being able to see the video, or hear the audio, or the line of communication dropping entirely. Despite this many are using screens on a daily basis for encounters that were previously not possible. This can prompt feelings of a co-presence but it can also be a poignant reminder of physicality when people cannot engage all the bodily senses they are used to engaging. Skype offers visuality, interactivity and simultaneity but it does replicate reality.

Ahmed (2006, 103) argues that: 'The intimacy of contact shapes bodies as they orientate toward each other doing different kinds of work'. Elizabeth Grosz (1995) makes reference to the body builder in relation to their weights, and the writer in relation to their pen and paper. Over the past few years, given the growth in many parts of the world in computers, laptops, iPads, mobile phones and other digital devices, it makes sense to consider orientations, proximities and melding in relation to bodies and machines (Mackenzie 2002).

> The work of inhabitance involves orientation devices; ways of extending bodies into spaces that create new folds, or new contours of what we could call liveable or inhabitable space. If orientation is about making the strange familiar through the extension of bodies into space, then dis-orientation occurs when that extension fails. Or we could say that some spaces extend certain bodies and simply do not leave room for others.
>
> (Ahmed 2006, 11)

Through Skype bodies are becoming 'reorientated'. Many, especially in the western world, no longer live just in material spaces but have learnt to engage with devices – 'hardware' and 'software' which transmit sound and image. These devices have extended bodies into 'spaces that create new folds' – new lines of sight/site, crossings and ways of engaging. Crossings, meetings and new orientations happen in a variety of online and offline spaces. The boundaries of the spaces we knew previously – our bedrooms, homes, hotels, workspaces, play spaces, buses and cars – have shifted in an increasingly globalized and connected world. More of us 'live apart together' (Holmes 2004) and binaries such as private/public, absent/present, close/distant are further destabilized (Cupples and Thompson 2010). Digital technologies are linking bodies and spaces in new ways, especially through the visual which enables us to read the bodily performances of those we are communicating with, but it also opens up possibilities to feel surveyed. In some instances interactions using Skype with video may result in feelings of happiness, being connected and reassured but in other instances they may result in feelings of frustration, intrusion and personal insecurity.

Back to writing tables and digital screens

In this final section I want to return to where I started, that is with a personal story and with Ahmed's reflections on tables – tables for writing, philosophising, computing and eating on. I suspect it is more than a coincidence that I wrote this book not at a writing table or desk but on a kitchen table. Ahmed (2006, 61) points out that there is a feminist publisher named Kitchen Table Press[3] and that the 'kitchen table provides the kind of surface on which women tend to work'. I have a perfectly good office at home, furnished with a desk, office chair, shelves, armchair to the side, rug on the floor and even a view. I also have a perfectly good office at my university, again well-appointed and with a view. And yet I am here! Sitting at the kitchen table. Why? This is where I feel orientated, able to write.

As I sat here typing my son arrived to make himself breakfast. The conversation was not scintillating but I welcomed the opportunity to stop typing, at least briefly and connect, at least at some level, with him. He will soon be back at university in another city and then we will have to Skype, phone, Snapchat and email to stay in touch. My two children are now in their twenties. I have been an academic their whole lives. This is a big part of why I have always worked at the kitchen table. It has been the centre of domestic life, where I, as 'Mum', have chosen to be simultaneously cooking, eating, nurturing and reading, preparing classes and writing. The kitchen table has then supported not just my domestic but also my political work. As a feminist geographer whose research has focused on pregnancy and mothering, the domestic and political have never been separate for me.

Interestingly, I now see reflecting on Ahmed's queer phenomenology that my body, its shape and orientation, is spatially manifest in the objects on the kitchen table. One end is covered in folders, books, papers, a pen, my glasses, laptop and a mouse; the other, a vase of flowers, ceramic plate and some serviettes. The scholarly and domestic are drawn together by a fabric runner down the middle. Ahmed (2006, 61) points out, kitchen tables are a 'reorientation device'. In reflecting now, maybe this book on Skype was always going to be a little queer given that I was writing it at the kitchen table. My other research projects, many of which were also written on kitchen tables, on things such as domestic bathrooms, fat bodies, men with breasts, and women 'coming out' as pregnant at work, to name just a few, have also been a little queer, I guess.

Last year when I took on a senior leadership role in my university, having been there 23 years, I shifted offices from geography to a larger, grander office in the central administration building. I wondered what books to put on my new shelves, what to adorn the walls with. I was arriving in a new space that did not feel like home. Did I have to hide certain titles from view? Could I write about bodies, queerness, sex and gender in this new office, at this new desk, surrounded by new objects with new people coming in and out? These questions are for another project but for now it is sufficient to make the point that:

For bodies to arrive in spaces where they are not already at home, where they are not 'in place,' involves hard work; indeed, it involves painstaking labour for bodies to inhabit spaces that do not extend their shape. Having arrived, such bodies in turn might acquire new shapes. And spaces in turn acquire new bodies. So, yes, we should celebrate new arrivals.

(Ahmed 2006, 62)

Skype is a new space, a new arrival, where bodies are taking time to feel 'in place'. For some, in some contexts, it has not yet extended its shape into the shape of the bodies using the platform. Technologically, socially and culturally some of us are not yet entirely sure how to occupy this space and yet we are learning. This is a new kind of work. Ahmed (2006, 100) argues that: 'The work of reorientation needs to be made visible as a form of work'.

This book has attempted to do exactly that; it has attempted to make visible how using Skype requires a reorientation – learning how to inhabit online and offline spaces differently, extending into them differently. I am certainly *not* arguing that Skype offers some kind of utopian future but I think that understanding its queer potential may help us see the value in 'mixing, or coming into contact with things that reside on different lines, as opening up new kinds of connection' (Ahmed 2006, 154–155). Skype offers opportunities such as being able to take up different sexed and gendered subjectivities that are non-normative, for grandparents and grandchildren to establish and maintain relationships of care that may earlier have been impossible, to take part in special occasions even if one is physically distant and to resist webcam if it feels intrusive and as though one is being surveilled. There is an opportunity to not simply replicate and reinforce existing power relations but filter them through a new frame with the hope that light can be cast in new directions. By understanding more about how Skype is disorientating and reorientating bodies and spaces, the hope is that it becomes possible to see more clearly new lines of sights/site, helping create a world in which many bodies, not just a privileged few, are able to 'sink' comfortably and feel supported. As Ahmed (2006, 158) astutely comments, it is not so much the disorientation that matters (because we all experience disorientation to some degree) but what we do with it and whether it can offer us 'the hope of new directions and whether new directions are reason enough for hope'. Herein lies our challenge.

Notes

1 In discussing 'the closet' Brown (2000, 147) also makes the important point that: 'People can be in and out of the closet simultaneously through resituating that broader environment' (also see Fuss 1991; Sedgwick 1990). Fuss (1991) explains the difficulty with the inside/outside rhetoric is that it disguises the fact that many people are both inside and outside simultaneously.
2 Elspeth Probyn in Blush: Faces of Shame (2005, 4) uses the concept of shame to: 'nudge readers to question their assumptions about the workings of our bodies and

their relations to thinking; about the nature of emotions in daily life and in academic reflection; and about ways of writing and relating'.

3   Kitchen Table: Women of Color Press was an activist press set up in 1989 by author Barbara Smith in Boston. In addition to publishing books aimed at promoting the writing of women of colour they were also an activist and advocacy organization devoted to the liberation struggles of all oppressed people. The suggestion for the group came originally from Smith's friend, poet Audre Lorde (Wikipedia: Kitchen Table: Women of Color Press 2015b).

Bibliography

#SkypeSavvy (2015). What moments will you make with skype? [Online]. Available at: http://moments.skype.com/skypesavvy/ [Accessed 21 February 2016].

Adams, P. (1998). Network topologies and virtual place. *Annals of the Association of American Geographers*, 88(1), pp. 88–106.

Adams, P. C. (1997). Cyberspace and virtual places. *Geographical Review*, 87(2), pp. 155–171.

Adams, P. C. (2007). Introduction to 'technological change': A special issue of ethics, place and environment. *Ethics, Place and Environment*, 10(1), pp. 1–6.

Adams, P. C. (2009). *Geographies of media and communication*. Malden: Wiley-Blackwell.

Adams, P. C. and Ghose, R. (2003). India.com: The construction of a space between. *Progress in Human Geography*, 27(4), pp. 414–437.

Adams-Hutcheson, G. and Longhurst, R. (under consideration). 'At least in person there would have been a cup of tea': Interviewing via Skype. *Area*.

Ahmed, S. (1998). *Differences that matter: Feminist theory and postmodernism*. Cambridge: Cambridge University Press.

Ahmed, S. (2000). *Strange encounters: Embodied others in post-coloniality*. London: Routledge.

Ahmed, S. (2004). *The cultural politics of emotion*. New York: Routledge.

Ahmed, S. (2006). *Queer phenomenology: Orientations, objects, others*. Durham and London: Duke University Press.

Ahmed, S. (2010a). *The promise of happiness*. Durham: Duke University Press.

Ahmed, S. (2010b). Feminist killjoys (and other willful Subjects). *Scholar and Feminist Online*, 8(3): Summer. [Online]. Available at: http://sfonline.barnard.edu/polyphonic/print_ahmed.htm [Accessed 3 November 2015].

Ahmed, S. (2012). *On being included: Racism and diversity in institutional life*. Durham: Duke University Press.

Ahmed, S. (2014). *Willful subjects*. Durham: Duke University Press.

Ahmed, S. (2016). Feminism and fragility. [Online]. Feministkilljoys - Sarah Ahmed Blog. Available at: http://feministkilljoys.com/2016/01/26/feminism-and-fragility/ [Accessed 1 February 2016].

Anderson, B. (2006). Becoming and being hopeful: Towards a theory of affect. *Environment and Planning D: Society and Space*, 24(5), pp. 733–752.

Anderson, B. and Harrison, P. (eds.) (2010). *Taking-place: Non-representational theories and geography*. Durham: Ashgate.

Anderson, B., Kearnes, M., McFarlane, C. and Swanton, D. (2012). On assemblages and geography. *Dialogues in Human Geography*, 2(2), pp. 171–189.

Anderson, K. (1995). Culture and nature at the Adelaide Zoo: At the frontiers of 'human' geography. *Transactions: Institute of British Geographers*, 20(3), pp. 275–294.

Anderson, K. and Smith, S. J. (2001). Editorial: Emotional geographies. *Transactions of the Institute of British Geographers*, 26(1), pp. 7–10.

Ash, J. (2009). Emerging spatialities of the screen: Video games and the reconfiguration of spatial awareness. *Environment and Planning A*, 41(9), pp. 2105–2124.

Ash, J. (2010). Architectures of affect: Anticipating and manipulating the event in the processes of videogame design and testing. *Environment and Planning D: Society and Space*, 28(4), pp. 653–671.

Ash, J. (2013). Rethinking affective atmospheres: Technology, perturbation and space times of the non-human. *Geoforum*, 49, pp. 20–28.

Ash, J. (2015a). Technology and affect: Towards a theory of inorganically organised objects. *Emotion, Space and Society*, 14(1), pp. 84–90.

Ash, J. (2015b). *The interface envelope: Gaming, technology, power* (1st ed.). New York: Bloomsbury Academic, an imprint of Bloomsbury Publishing, Inc.

Ash, J. and Simpson, P. (2014). Geography and post-phenomenology. *Progress in Human Geography*, 40(1), pp. 48–66.

Bailo, P.J. (2013). *The essential digital interview handbook: Lights, camera, interview: Tips for skype, google hangout, gotomeeting, and more*. Pompton Plains: Career Press.

Bain, A.L., and Nash, C.J. (2006). Undressing the researcher: Feminism, embodiment and sexuality at a queer bathhouse event. *Area*, 38(1), pp. 99–106.

Baker, A. (2000). Two by two in cyberspace: Getting together and connecting online. *CyberPsychology and Behavior*, 3(2), pp. 237–242.

Baker, A. J. (2002). What makes an online relationship successful? Clues from couples who met in cyberspace. *CyberPsychology and Behavior*, 5(4), pp. 363–375.

Baker, A. J. (2005). Double click: Romance and commitment among online couples. Cresskill: Hampton Press.

Baker, A. J. (2008). Down the rabbit hole: The role of place in the initiation and development of online relationships. In: A. Barak, ed., *Psychological aspects of cyberspace: Theory, research, applications*. Cambridge: Cambridge University Press, pp. 163–184.

Baker, A. J. and Whitty, M. T. (2008). Researching romance and sexuality online: Issues for new and current researchers. In: S. Holland, ed., *Remote relationships in a small world*. New York: Peter Lang, pp. 34–49.

Barad, K. (2003). Post humanist performativity: Toward an understanding of how matter comes to matter. *Signs: Journal of Women in Culture and Society*, 28(3), pp. 801–832.

Barraket, J. and Henry-Waring, M. (2006). Online dating and intimacy in a mobile world. In: V. Colic-Peisker, B. McNamara and F. Tilbury, eds., *Sociology for a mobile world: Proceedings of the Australian Sociological Association 2006 Conference*. Australia: The Australian Sociological Association (TASA), University of Western Australia and Murdoch University, pp. 1–10.

Batinic, B., Reips, U.D. and Bosnjak, M. (eds.) (2002). *Online social sciences*. Seattle: Hogrefe and Huber Publishers.

Baym, N. (2010). *Personal connections in the digital age*. Cambridge: Polity.

Bell, A., Crothers, C., Goodwin, I., Kripalani, K., Sherman, K. and Smith, P. (2008). *World internet project New Zealand: The internet in New Zealand 2007, final report*.

2007 Benchmark Survey. Auckland: The Institute of Culture, Discourse and Communication and AUT University.

Bell, D. (2009). Cyberspace/cyberculture. In: R. Kitchin and N. Thrift, eds., *International Encyclopaedia of Human Geography*, 3. Oxford: Elsevier, pp. 468–472.

Bell, D., Binnie, J., Cream, J. and Valentine, G. (1994). All hyped up and no place to go. *Gender, Place and Culture*, 1(1), pp. 31–47.

Bell, D. and Kennedy, B.M. (2007). *The cyberculture reader* (Second ed.). London: Routledge.

Ben-Ze'ev, A. (2004). *Love on-line: Emotions on the internet.* Cambridge: Cambridge University Press.

Benski, F. and Fisher, E. (eds.) (2013). *Internet and emotions.* London: Routledge.

Berg, L. and Kearns, R.A. (1996). Naming as norming: 'Race', gender and the identity politics in Aotearoa/New Zealand. *Environment and Planning D: Society and Space*, 14(1), pp. 99–122.

Bingham, N. (1996). Object-ions: From technological determinism towards geographies of relations. *Environment and Planning D: Society and Space*, 14(6), pp. 635–657.

Bingham, N. (2005). Socio-technical. In: D. Atkinson, P. Jackson, D. Sibley, and N. Washbourne, eds., *Cultural geography.* London: IB Tauris, pp. 200–206.

Blunt, A. (2005). Cultural geography: Cultural geographies of home. *Progress in Human Geography*, 29(4), pp. 505–515.

Blunt, A. and Dowling, R. (2006). *Home.* New York: Routledge.

Bogle, J. (2014). *Time and space are no big deal, thanks to skype.* [Online]. Skype Blogs. Available at: http://blogs.skype.com/2014/02/28/time-and-space-are-no-big-deal-tha nks-to-skype/ [Accessed 2 May 2014].

Bondi, L. (2003). Empathy and identification: Conceptual resources for feminist fieldwork. *ACME*, 2(1), pp. 64–76.

Bondi, L. (2005). Making connections and thinking through emotions: Between geography and psychotherapy. *Transactions: Institute of British Geographers*, 30(4), pp. 433–448.

Bondi, L. (2014). Understanding feelings: Engaging with unconscious communication and embodied knowledge. *Emotion, Space and Society*, 10, pp. 44–54.

Bondi, L. and Davidson, J. (2011). Lost in translation. *Transactions of the Institute of British Geographers*, 36, pp. 595–598.

Braziel, J. E. and LeBesco, K. (2001). *Bodies out of bounds: Fatness and transgression.* Berkeley: University of California Press.

Brennan, T. (2004). *The transmission of affect.* Ithaca: Cornell University Press.

Brown, A. (2012). 'No one-handed typing': An exploration of gameness, rules and spoilsports in an erotic role play community in world of warcraft. *Journal of Gaming and Virtual Worlds*, 4(3), pp. 259–273.

Brown, G., Browne, K., Brown, M., Roelvink, G., Carnegie, M. and Anderson, B. (2011). Sedgwick's geographies: Touching space. *Progress in Human Geography*, 35(1), pp. 121–131.

Brown, M. (2000). *Closet space: Geographies of metaphor from the body to the globe.* London: Routledge.

Bryson, M. (2004). When Jill Jacks in: Queer women and the net. *Feminist Media Studies*, 4(3), pp. 239–254.

Buchanan, T. and Whitty, M.T. (2013). The online dating romance scam: Causes and consequences of victimhood. *Psychology, Crime and Law*, 20(3), pp. 261–283.

Burns, E. (2010). Developing email interview practices in qualitative research. *Socio-logical Research Online*, 15(4). [Online]. Available at: http://www.socresonline.org.uk/15/4/8.html [Accessed 4 February 2016].

Butler, J. (1990). *Gender trouble: Feminism and the subversion of identity*. New York: Routledge.

Butler, J. (1993). *Bodies that matter: On the discursive limits of 'sex'*. New York: Routledge.

Butler, R. and Bowlby, S. (1997). Bodies and spaces: An exploration of disabled people's experiences of public space. *Environment and Planning D: Society and Space*, 15(4), pp. 411–433.

Butler, R. and Parr, H. (eds.) (1999). *Mind and body spaces: Geographies of illness, impairment, and disability*. London: Routledge.

Buttimer, A. (1979). Reason, rationality and human creativity, *Geografiska Annaler*, 61B, pp. 43–49.

Cassidy, M. F. (2001). Cyberspace meets domestic space: Personal computers, women's work, and the gendered territories of the family home. *Critical Studies in Media Communication*, 18(1), pp. 44–65.

Caukin, J. (2011). Your #ILOVESKYPE stories, play blog. [Online]. Skype Website. Available at: http://blogs.skype.com/2011/08/30/your-stories/ [Accessed 14 February 2016].

Chen, P. and Hinton, S. (1999). Real time interviewing using the world wide web . *Sociological Research Online*, 4(3). [Online]. Available at: http://socresonline.org.uk/4/3/chen.html [Accessed 14 February 2016].

Chow-White, P. A. (2006). Race, gender and sex on the net: Semantic networks of selling and storytelling sex tourism. *Media, Culture and Society*, 28(6), pp. 883–905.

Clough, P. T. (2000). *Autoaffection: Unconscious thought in the age of technology*. Minneapolis: University of Minnesota Press.

Colls, R. (2006). Outsize/outside: Bodily bigness and the emotional experiences of British women shopping for clothes. *Gender, Place and Culture*, 13(5), pp. 529–545.

Colls, R. (2007). Materialising bodily matter: Intra-action and the embodiment of 'fat'. *Geoforum*, 38(2), pp. 353–365.

Cooper, A. and Sportolari, L. (1997). Romance in cyberspace: Understanding online attraction. *Journal of Sex Education and Therapy*, 22(1), pp. 7–14.

Cornwell, B. and Lundgren, D. C. (2001). Love on the internet: Involvement and misrepresentation in romantic relationships in cyberspace vs. realspace. *Computers in Human Behaviour*, 17(2), pp. 197–211.

Correa, T. (2014). Bottom-up technology transmission within families: Exploring how youths influence their parents' digital media use with dyadic data. *Journal of Communication*, 64(1), pp. 103–124.

Couldry, N. and McCarthy, A. (eds.) (2004) *Mediaspace: Place, scale and culture in a media age*. London: Routledge.

Crang, M., Crang, P. and May, J. (1999). *Virtual geographies: Bodies, spaces and relations*. London: Routledge.

Cranny-Francis, A. (2013). *Technology and touch: The biopolitics of emerging technologies*. Palgrave Macmillan: New York.

Cupples, J. and Thompson, L. (2010). Heterotextuality and digital foreplay: Cell phones and the culture of teenage romance. *Feminist Media Studies*, 10(1) pp. 1–17.

Davidson, J. (2001). Fear and trembling in the mall: Women, agoraphobia, and body boundaries. In: I. Dyck, N. Lewis, and S. McLafferty, eds., *Geographies of women's Health*, 1st edition. New York: Routledge, pp. 213–230.

Davidson, J. (2003). *Phobic geographies: The phenomenology and spatiality of identity.* Aldershot: Ashgate.

Davidson, J. (2008). Autistic culture online: Virtual communication and cultural expression on the spectrum. *Social and Cultural Geography*, 9(7), pp. 791–806.

Davidson, J. and Bondi, L. (2004). Spatialising affect; affecting space: Introducing emotional geographies. *Gender, Place and Culture*, 11(3), pp. 373–374.

Davidson, J. and Milligan, C. (2004). Embodying emotion sensing space: Introducing emotional geographies. *Social and Cultural Geography*, 5(4), pp. 523–532.

Davidson, J. and Parr, H. (2010). Enabling cultures of dis/order online. In: V. Chouinard, E. Hall, and R. Wilton, eds., *Towards enabling geographies: 'Disabled' bodies and minds in society and space.* Farnham: Ashgate, pp. 63–83.

Davidson, J.Bondi, L. and Smith, M. (eds.) (2005). *Emotional geographies.* Farnham: Ashgate.

Davis, K. (ed.) (1997). *Embodied practices: Feminist perspectives on the body.* London: Sage.

De Beer, M. (2014). The skype community welcomes its 2,000,000th user. [Online]. Skype Blogs. Available at:http://blogs.skype.com/2014/05/01/the-skype-community-welcomes-its-2000000th-user/ [Accessed 2 May 2014].

De Jong, A. (2015). Using Facebook as a space for storytelling in geographical research. *Geographical Research*, 53(2), pp. 211–223.

Deakin, H. and Wakefield, K. (2013). Skype interviewing: Reflections of two PhD researchers. *Qualitative Research*, 14(5), pp. 603–616.

Del Casino, V. and Brooks, C. (2014). Talking about bodies online: Viagra, YouTube, and the politics of public(ized) sexualities. *Gender, Place and Culture*, 22(4), pp. 474–493.

Derrida, J. (1988). *Limited inc.* In: G. Gerald, ed., Evanston: Northwestern University Press.

Dixon, D. and Straughan, E. (2010). Geographies of touch/touched by geography. *Geography Compass*, 4(5), pp. 449–459.

Dockweiler, S. (2016). The new secrets to rocking your skype interview. [Online]. The Muse. Available at: https://www.themuse.com/advice/the-new-secrets-to-rocking-your-skype-interview [Accessed 10 January 2016].

Dodge, M. and Kitchen, R. (2001). *Mapping cyberspace.* London: Routledge.

Domosh, M. (1998). Geography and gender: Home, again? *Progress in Human Geography*, 22(2), pp. 276–282.

Döring, N. (2000). Feminist views of cybersex: Victimization, liberation, and empowerment. *CyberPsychology* and *Behavior*, 3(5), pp. 863–884.

Döring, N. (2002). Studying online-love and cyber-romance. In B. Batinic, U.D. Reips, and M. Bosnjak, eds., *On-line social sciences.* Germany: Hogrefe and Huber Publishers, pp. 333–356.

Driscoll, C. and Greg, M. (2010). My profile: The ethics of virtual ethnography. *Emotion, Space and Society*, 3, pp. 15–20.

Dumitrica, D. and Wyatt, S. (2015). Guest editorial: Digital technologies and Social transformations: What role for critical theory. *Canadian Journal of Communcation*, 40(4), pp. 589–596.

Duncan, N. (ed.) (1996). *Bodyspace: Destabilizing geographies of gender and sexuality.* London: Routledge.

Edensor, T. (2010). Geographies of rhythm: Nature, place, mobilities and bodies. Farnham: Ashgate.

Ekman, P. (1993). Facial expression and emotion. *American Psychologist*, 48(4), pp. 384–392.

Farman, J. (2012). *Mobile interface theory: Embodied space and locative media*. New York: Routledge.

Feenberg, A. (1991). *Critical theory of technology*. New York: Oxford University Press.

Feenberg, A. (1999). *Questioning technology*. London: Routledge.

Flanagan, M. (2000). Navigating the narrative in space: Gender and spatiality in virtual worlds. *Art Journal*, 59(3), pp. 74–85.

Fluri, J.L. (2006). 'Our website was revolutionary': Virtual spaces of representation and resistance. *ACME, Special Issue on Gender, Space and Technology*, 5(1), pp. 89–111.

Fortunati, L. and Taipale, S. (2012). Women's emotions towards the mobile phone. *Feminist Media Studies*, 12(4), pp. 538–549.

Foucault, M. (1980). *Power/knowledge: Selected interviews and other writings 1972–1977*. New York: Pantheon Books.

Friedman, M. and Calixte, S. (eds.) (2009). *Mothering and blogging: The radical act of mommyblog*. Toronto: Demeter Press.

Frizzo-Barker, J. and Chow-White, P. A. (2012). 'There's an App for that': Mediating mobile moms and connected careerists through smartphones and networked individualism. *Feminist Media Studies*, 12(4), pp. 580–589.

Frost, J., Chance, Z., Norton, M. and Ariely, D. (2008). People are experience goods: Improving online dating with virtual dates. *Journal of Interactive Marketing*, 22(1), pp. 51–61.

Fuchs, C. (2009a). Information and communication technologies and society: A contribution to the critique of the political economy of the Internet. *European Journal of Communication*, 24(1), pp. 69–87.

Fuchs, C. (2009b). A contribution to theoretical foundations of critical media and communication studies. *Javnost – The Public*, 16(2), pp. 5–24.

Fuss, D. (2004). *The sense of an interior: Four writers and the rooms that shaped them*. New York: Routledge.

Fuss, D. (ed.) (1991). *Inside/out: Lesbian theories, gay theories*. New York: Routledge.

Gatens, M. (1988). Towards a feminist philosophy of the body. In: B. Caine, E. Grosz, and M. de Lepervanche, eds., *Crossing boundaries: Feminisms and critiques of knowledges*. Sydney: Allen and Unwin, pp. 59–70.

Gatens, M. (1991). Corporeal representation in/and the body politic. In R. Diprose and R. Ferrell, eds., *Cartographies: Poststructuralism and the mapping of bodies and spaces*. Sydney: Allen and Unwin, pp. 79–87.

Gatrell, C. (2008). *Embodying women's work*. England: Open University Press.

Gibson, P. C. (ed.) (2004). *More dirty looks: Gender, pornography and power*. London: British Film Industry.

Gibson, A., Miller, M., Smith, P., Bell, A. and Crothers, C. (2013). World Internet Project New Zealand. *The Internet in New Zealand 2013*, Auckland: Institute of Culture, Discourse & Communication, AUT University. Available at: http://www.aut.ac.nz/data/assets/pdf_file/0007/424816/wipnz2013final.pdf

Gilbert, R. L., Murphy, N. A. and Avalos, M. C. (2011). Communication patterns and satisfaction levels in three-dimensional versus real-life intimate relationships. *CyberPsychology, Behavior and Social Networking*, 14(10), pp. 585–589.

Giles, F. (2004). 'Relational, and strange': A preliminary foray into a project to queer breastfeeding. *Australian Feminist Studies*, 19(45), pp. 302–315.

Goorwich, S. (2015). This vibrator has an inbuilts camera and even syncs with FaceTime. [Online]. Metro. Available at: http://metro.co.uk/2015/02/28/this-vibrator-has-an-inbuilt-camera-and-even-syncs-with-facetime-5083198/ [Accessed 26 March 2015].

Gorton, K. (2007). Theorizing emotion and affect: Feminist engagements . *Feminist Theory*, 8(3), pp. 333–348.

Graham, S. (1998). The end of geography or the explosion of place? Conceptualising space, place and information technologies. *Progress in Human Geography*, 22(2), pp. 165–185.

Gregson, N. and Rose, G. (2000). Taking Butler elsewhere: Performativities, spatialities and subjectivities. *Environment and Planning D: Society and Space*, 18(4), pp. 433–452.

Grosz, E. (1990). The body of signification. In J. Fletcher and A. Benjamin, eds., *Abjection, melancholia and love: The work of Julia Kristeva*. London: Routledge, pp. 80–103.

Grosz, E. A. (1989). *Sexual subversions: Three French feminists*. Sydney: Allen and Unwin.

Grosz, E. A. (1994). *Volatile bodies: Toward a corporeal feminism*. St Leonards: Allen and Unwin.

Grosz, E. A. (1995). *Space, time and perversion*. St. Leonards: Allen and Unwin.

Hanna, P. (2012). Using internet technologies (such as Skype) as a research medium: A research note. *Qualitative Research*, 12(2), pp. 239–242.

Haraway, D. (1990). A cyborg manifesto: Science, technology, and socialist-feminism in the late twentieth century. In L. J. Nicholson ed., *Feminism/Postmodernism*, New York: Routledge, pp. 190–233.

Haraway, D. (1991). *Simians, cyborgs and women: The reinvention of nature*. New York: Routledge.

Hardey, M. (2002). Life beyond the screen: Embodiment and identity through the Internet. *The Sociological Review*, 50(4), pp. 570–585.

Helsper, E. J. and Whitty, M. T. (2010). Netiquette within married couples: Agreement about acceptable online behavior and surveillance between partners. *Computers in Human Behavior*, 26(5), pp. 916–926.

Hertlein, K. and Sendak, S. (2007). *Love 'bytes': Internet infidelity and the meaning of intimacy in computer-mediated relationships*. [Online]. Inter-disciplinary.net. Available at: www.inter-disciplinary.net/ptb/persons/pil/pil1/hertleinsendak%20paper.pdf [Accessed 26 November 2015].

Hillis, K. (1996). A geography of the eye: The technologies of virtual reality. In: R. Shields, ed., *Cultures of Internet: Virtual spaces, real histories, living bodies*. London: Sage, 70–98.

Hillis, K. (1999). *Digital sensations: Space, identity, and embodiment in virtual reality*. NED – New edition, Minneapolis: University of Minnesota Press.

Hillis, K., Paasonen, S. and Petit, M. (eds.) (2015). *Networked affect*. Cambridge: The MIT Press.

Holloway, S. L. and Valentine, G. (2001a). 'It's only as stupid as you are': Children's and adults' negotiation of ICT competence at home and at school. *Social and Cultural Geography*, 2(1), pp. 25–42.

Holloway, S. L. and Valentine, G. (2001b). Children at home in the wired world: Reshaping and rethinking the home in urban geography. *Urban Geography*, 22(6), pp. 562–583.

Holloway, S. L. and Valentine, G. (2001c). Placing cyberspace: Processes of Americanization in British children's use of the internet. *Area*, 33(2), pp. 153–160.

Holmes, M. (2004). An equal distance?: Individualisation, gender, and intimacy in distance relationships. *The Sociological Review*, 52(2), pp. 180–200.

hooks, B. (1991). Homeplace: A site of resistance, excerpts from Yearning: Race, gender and cultural politics. In: L. McDowell, ed., *Undoing place? A geographical Reader*. London: Arnold, pp. 33–38.

Hopkins, P. and Pain, R. (2007). Geographies of age: Thinking relationally. *Area*, 39(3), pp. 287–294.

Hopkins, P., Olson, E., Pain, R. and Vincett, G. (2010). Mapping intergenerationalities: The formation of youthful religiosities. *Transactions of the Institute of British Geographers*, 36(2), pp. 314–327.

Horst, H. A. (2009). Aesthetics of the self digital mediations. In: D. Miller, ed., *Anthropology and the individual*. Oxford: Berg, pp. 99–113.

Horst, H. A. and Miller, D. (2006). *The cell phone: An anthropology of communication*. Oxford: Berg.

Horst, H. A. and Miller, D. (eds.) (2012). *Digital anthropology*. Oxford: Berg.

Husserl, E. (1989). *Ideas pertaining to a pure phenomenology and to a phenomenological philosophy, Second Book*, translated by R. Rojcewicz and A. Schuwer. Dordrecht: Kluwer Academic Publishers.

Hutcheson, G. (2013). Methodological reflections on transference and counter-transference in geographical research: Relocation experiences from post-disaster Christchurch, Aotearoa New Zealand. *Area*, 45(4), pp. 477–484.

International Telecommunications Union (2014). Measuring the information society report 2014, Place des Nations CH-1211 Geneva Switzerland. Available at: https://www.itu.int/en/ITU-D/Statistics/Documents/publications/mis2014/MIS2014_with out_Annex_4.pdf [Accessed 6 March 2016].

Jagose, A. (1996). *Queer theory*. Melbourne: Melbourne University Press.

Janelle, D. and Hodge, D. (eds.) (2000). *Information, place, and cyberspace: Issues in accessibility*. Berlin: Springer.

Jay, M. (1993). *Downcast eyes: The denigration of vision in twentieth-century French thought*. Berkeley: University of California Press.

Johnson, L. (1989). Embodying geography–Some implications of considering the sexed body in space. In: New Zealand Geographical Society Proceedings of the 15th New Zealand Geography Conference. Dunedin, New Zealand, August, pp. 134–138.

Johnston, L. (1997). Queen(s') street or Ponsonby poofters? The embodied HERO parade sites. *New Zealand Geographer*, 53(2), pp. 29–33.

Johnston, L. (2005). *Queering tourism: Paradoxical performances at gay pride parades*. London: Routledge.

Johnston, L. (2012). Sites of excess: The spatial politics of touch for drag queens in Aotearoa New Zealand. *Emotion, Space and Society* 5(1), pp. 1–9.

Johnston, L. (2015). Gender and sexuality I: Genderqueer geographies? *Progress in Human Geography*, published online before print June 19, 2015, doi:10.1177/0309132515592109.

Johnston, L. and Longhurst, R. (2010). *Space, place and sex: Geographies of sexualities*. Lanham: Rowman and Littlefield.

Johnston, L. and Valentine, G. (1995). Wherever I lay my girlfriend, that's my home: The performance and surveillance of lesbian identities in domestic environments. In: D. Bell and G. Valentine, eds., *Mapping desire: Geographies of sexualities*. London: Routledge, pp. 99–113.

Jones, S. (ed.) (1999). *Doing internet research: Critical issues and methods for examining the net*. California: Sage.

Kang, T. (2012). Gendered media, changing intimacy: Internet-mediated transnational communication in the family sphere. *Media, Culture* and *Society*, 34(2), pp. 146–161.

Karatzogianni, A. and KunstmanA. (eds.) (2012). Digital cultures and the politics of emotion: Feelings, affect and technological change. Basingstoke: Palgrave Macmillan.

Kingsbury, P. and Pile, S. (eds.) (2014). *Psychoanalytic geographies*. England: Ashgate.

Kinsley, S. (2013a). The matter of 'virtual' geographies. *Progress in Human Geography*, 38(3), pp. 364–384.

Kinsley, S. (2013b). Beyond the screen: Methods for investigating geographies of life 'online'. *Geography Compass*, 7(8), pp. 540–555.

Kinsley, S. (2016) Vulgar geographies? Popular cultural geographies and technology. *Social and Cultural Geography*, in press. [Online]. Available at: https://www.resea rchgate.net/publication/287633106_Vulgar_Geographies_Popular_Cultural_Geogra phies_and_Technology [Accessed 26 February 2016].

Kitchin, R. (1998a). Towards geographies of cyberspace. *Progress in Human Geography*, 22(3), pp. 385–406.

Kitchin, R. (1998b). *Cyberspace: The world in the wires*. London: John Wiley.

Kitchin, R. (2014). *The data revolution: big data, open data, data infrastructures and their consequences*. London: Sage.

Kitchin, R. and Dodge, M. (2011). *Code/space: Software and everyday life*. Cambridge: MIT Press.

Koerber, A. (2001). Postmodern, resistance, and cyberspace: Making rhetorical spaces for feminist mothers on the web. *Women's Studies in Communication*, 24(2), pp. 218–240.

Kolotkin, R. A., Williams, M. M., Lloyd, C. D. and Hallford, E. W. (2012). Does loving an avatar threaten real life marriage? *Journal of Virtual Worlds Research*, 5(3), pp. 1–40.

Lalibertè, N. and Schurr, C. (2015). The stickiness of emotions in the field: Complicating feminist methodologies. *Gender, Place* and *Culture*, pp. 1–7.

Larner, W. and Spoonley, P. (1995). Postcolonial politics in Aotearoa/New Zealand. In: D. Stasiulis and N. Yuval-Davis, ed., *Unsettling settler societies: Articulations of gender, race, ethnicity and class*. London: Sage, pp. 39–64.

Latour, B. (1993). *We have never been modern*. Translated by D. Nicholson-Smith. Oxford: Blackwell.

Latour, B. (1999). *Pandora's hope: essays on the reality of science studies*. Cambridge: Harvard University Press.

Lea, M. and Spears, R. (1995). *Love at first byte? Building personal relationships over computer networks*. Thousand Oaks: Sage.

Lefebvre, H. (1991). *The production of space*, translated by D. Nicholson-Smith. Oxford: Blackwell.

Lefebvre, H. (2004). *Rhythmananalyis: Space, time and everyday life*, translated by S. Elden, and G. Moore. London: Continuum.

Lim, S. S. (2008). Technology domestication in the Asian homestead: Comparing the experiences of middle class families in China and South Korea. *East Asian Science, Technology* and *Society*, 2(2), pp. 189–209.

Lim, S. S. and Soon, C. (2010). The influence of social and cultural factors on mothers' domestication of household ICTs–Experiences of Chinese and Korean women. *Telematics and Informatics*, 27(3), pp. 205–216.

Longhurst, R. (2001). *Bodies: Exploring fluid boundaries.* London: Routledge.

Longhurst, R. (2008). *Maternities: Gender, bodies and space.* London: Routledge.

Longhurst, R. (2009). YouTube: A new space for birth? *Feminist Review*, 93(1), pp. 46–63.

Longhurst, R. (2013). Using Skype to mother: Bodies, emotions, visuality and screen. *Environment and Planning D: Society and Space*, 31(4), pp. 664–679.

Longhurst, R. (2014). Queering body size and shape: Performativity, the closet, shame, and orientation. In: C. Pausé, J. Wykes, and S. Murray eds., *Queering fat embodiment*, Farnham: Ashgate, pp13–25.

Longhurst, R. (2015). Mothering, digital media and emotional geographies in Hamilton, Aotearoa New Zealand. *Social and Cultural Geography*, 17(1), pp. 120–139.

Longhurst, R. and Johnston, L. (2014). Bodies, gender, place and culture: 21 years on. *Gender, Place and Culture*, 21(3), pp. 267–278.

Longhurst, R., Ho, E. and Johnston, L. (2008). Using 'the body' as an 'instrument of research': Kimch'i and Pavlova. *Area*, 40(2), pp. 208–217.

Longhurst, R., Johnston, L. and Ho, E. (2009). A visceral approach: Cooking 'at home' with migrant women in Hamilton, New Zealand. *Transactions of the Institute of British Geographers*, 34(3), pp. 333–345.

Lorimer, H. (2008). Cultural geography: Non-representational conditions and concerns. *Progress in Human Geography*, 32(4), 551–559.

Mackenzie, A. (2002). *Transductions: Bodies and machines at speed*, 1st edn, New York, Continuum.

MacKian, S. (2004). Mapping reflexive communities: Visualizing the geographies of emotion. *Social* and *Cultural Geography*, 5(4), pp. 615–631.

Madge, C. (2010). Internet mediated research. In: N. Clifford, S. French, and G. Valentine, eds., *Key methods in geography.* London: Sage, pp. 173–188.

Madge, C. and O'Connor, H. (2002). Online with e-mums: Exploring the Internet as a medium for research. *Area*, 34(1), pp. 92–102.

Madge, C. and O'Connor, H. (2005). Mothers in the making? Exploring liminality in cyber/space. *Transactions of the Institute of British Geographers*, 30(1), pp. 83–97.

Madianou, M. and Miller, D. (2011). Mobile phone parenting: Reconfiguring relationships between Filipina mothers and their left behind children. *New Media and Society*, 13(3), pp. 457–470.

Madianou, M. and Miller, D. (2012). *Migration and new media: Transnational families and polymedia.* London: Routledge.

Maginn, P. and Steinmetz, C. (eds.) (2015). *(Sub)urban sexscapes: Geographies and regulation of the sex industry.* London: Routledge.

Mann, R.Tarrant, A. and Leeson, G. W. (2015). Grandfatherhood: Shifting masculinities in later life. *Sociology*, pp. 1–17.

Massey, D. (1996). Masculinity, dualisms and high technology. *Transactions of the Institute of British Geographers*, 20(4), pp. 487–499.

Massumi, B. (2002). *Parables for the virtual: Movement, affect, sensation.* Durham: Duke University Press.

May, G. (2014). Yes, I f***** my iPad: Introducing the new adult toy that lets you have sex with your iPad. [Online]. Available at: http://metro.co.uk/2014/07/11/yes-i-fcked-my-ipad-a-probing-review-of-the-new-adult-toy-that-lets-you-have-sex-with-your-ipad-4794617 [Accessed 26 March 2015].

McDowell, L. and Court, G. (1994). Performing work: Bodily representations in merchant banks. *Environment and Planning D: Society and Space*12, pp. 727–750.

McKee, A.McNair, B. and Watson, A.F. (2015). Sex and the virtual suburbs: the pornosphere and community standards. In: P.J. Maginn, and C. Steinmetz, eds., *(Sub) urban sexscapes: Geographies and regulation of the sex industry*. London: Routledge.

McKenna, K. Y. A., Green, A. S. and Gleason, M. E. J. (2002). Relationship formation on the internet: What's the big attraction? *Journal of Social Issues*, 58(1), pp. 9–31.

McLelland, M. (2002). The new half net: Japan's 'intermediate sex' on-line. *International Journal of Sexuality and Gender Studies*, 7(2), pp. 163–175.

Meikle, G. (2016) *Social media: Communication, sharing and visibility*, London: Routledge.

Meunier, D. (2010). Emotional encounter: Researching young people's intimate spaces in their information and communication technology practices. *Emotion, Space and Society*, 3(1), pp. 36–39.

Miller, D. (2011). *Tales from Facebook*. Cambridge: Polity.

Miller, D. and Sinanan, J. (2014). *Webcam*. Cambridge: Polity.

Modern Family (2015). *American Skyper*, S6, E24, American Television Mockumentary created by S. Levitan and C. Lloyd, premiered on ABC.

Morley, D. (2002). *Home territories: Media, mobility, identity*. London: Routledge.

Morrison, C. (2012). Heterosexuality and home: Intimacies of space and spaces of touch. *Emotion, Space and Society*, 5(1), pp. 10–18.

Moss, P. (2001). Writing one's life. In: P. Moss, ed., *Placing autobiography in geography*. Syracuse: Syracuse University Press, pp. 1–21.

Moss, P. and Dyck, I. (1996). Inquiry into environment and body: Women, work and chronic illness. *Environment and Planning D: Society and Space*, 14(6), pp. 737–753.

Moss, P. and Dyck, I. (2002). *Women, body, illness: Space and identity in the everyday lives of women with chronic illness*. Lanham: Rowman and Littlefield.

Murnen, S.K. and Seabrook, R. (2012). Feminist perspectives on body image and physical appearance. In: T.F. Cash ,ed., *Encyclopaedia of body image and physical appearance volume 1*. San Diego: Academic Press, pp. 438–443.

Nast, H. and Pile, S. (eds.) (1998). *Places through the body*. London: Routledge.

Nayar, P. K. (2010). *An introduction to new media and cybercultures*. Malden: Wiley-Blackwell.

Nelson, L. (1999). Bodies (and spaces) do matter: The limits of performativity. *Gender, Place and Culture*, 6(4), pp. 331–353.

O'Connor, H., Madge, C., Shaw, R. and Wellens, J. (2008). Internet based interviewing. In: R. Lee, N. Fielding and G. Blank, eds., *The Sage handbook of online research methods*. London: Sage, pp. 271–289.

Opdenakker, R. (2006). Advantages and disadvantages of four interview techniques in qualitative research. *FQS Forum: Qualitative Social Research Sozialforschung*, 7(4), Art. 11. [Online]. Available at: http://www.qualitative-research.net/index.php/fqs/article/view/175/391 [Accessed 26 March 2015].

Oswin, N. (2008). Critical geographies and the uses of sexuality: Deconstructing queer space. *Progress in Human Geography*, 32(1), pp. 89–103.

Parks, M. R. and Roberts, L. D. (1998). 'Making moosic': The development of personal relationships on line and a comparison to their off-line counterparts. *Journal of Social and Personal Relationships*, 15(4), pp. 517–537.

Parr, H. (2002). New body-geographies: The embodied spaces of health and medical information on the internet. *Environment and Planning D: Society and Space*, 20(1), pp. 73–95.

Parr, H. (2005). Emotional geographies. In: P. Cloke, P. Crang and M. Goodwin, eds., *Introducing human geographies*, 2nd edition. London: Arnold, pp. 472–484.

Paterson, M. (2006). Feel the presence: Technologies of touch and distance. *Environment and Planning D: Society and Space*, 24(5) pp. 691–708.

Paterson, M. (2007). *The senses of touch: Haptics, affects, and technologies*, 1st edn. New York: Berg.

Paterson, M. (2009). Haptic geographies: Ethnography, haptic knowledges and sensuous dispositions. *Progress in Human Geography*, 33(6), pp. 766–788.

Paterson, M. and Dodge, M. (2012). *Touching space, placing touch*. Farnham: Ashgate.

Pausé, C., Wykes, J. and Murray, S. (eds.) (2014). *Queering fat embodiment*. Farnham: Ashgate.

Pile, S. (1996). *The body and the city: Psychoanalysis, space and subjectivity*. London: Routledge.

Pile, S. (2010a). Emotions and affect in recent human geography. *Transactions of the Institute of British Geographers*, 35(1), pp. 5–20.

Pile, S. (2010b). Intimate distance: The unconscious dimensions of the rapport between researcher and researched. *The Professional Geographer*, 62(4), pp. 483–495.

Pile, S. (2011). For a geographical understanding of affect and emotions. *Transactions of the Institute of British Geographers*, 36(4), pp. 603–606.

Porteous, J. D. (1986). Intimate sensing. *Area*, 18(3), pp. 250–251.

Pratt, G. in collaboration with the Philippine Women Centre, Vancouver, Canada (1998) Inscribing domestic work on Filipina bodies. In: H. J. Hast and S. Pile, eds., *Places through the body*. London: Routledge.

Probyn, E. (2000). *Carnal appetites: Foodsexidentities*. London: Routledge.

Probyn, E. (2003). The spatial imperative of subjectivity. In: K. Anderson, M. Domosh, S. Pile and N. Thrift, eds., *Handbook of cultural geography*. London: Sage, pp. 290–299.

Probyn, E. (2005). *Blush: Faces of shame*. Minneapolis: University of Minnesota Press.

Radde-Antweiler, K. (2007). Cyber-rituals in virtual worlds: Wedding-online in second life. *Masaryk University Journal of Law and Technology*, 1(2), pp. 185–196.

Ragnedda, M. and Muschert, G.W. (eds) (2013). *The digital divide: The internet and social inequality in international perspective*. Oxon: Routledge.

Rakow, L. F. and Navarro, V. (1993). Remote mothering and the parallel shift: Women meet the cellular telephone. *Critical Studies in Mass Communication*, 10(2), pp. 144–157.

Rich, A. (1986). Notes towards a politics of location. In: A. Rich, *Blood, bread and poetry: Selected prose 1979-1985*. New York: W. W. Norton and Company, pp. 210–232.

Roberts, E. (2012). Geography and the visual image: A huntological approach. *Progress in Human Geography*, 37(3), pp. 386–402.

Rodaway, P. (1994). *Sensuous geographies: Body, sense and place*. London: Routledge.

Rose, G. (1997). Situating knowledges: Positionality, reflexivities and other tactics. *Progress in Human Geography*, 21, pp. 305–320.

Rose, G. (1993). *Feminism and geography: The limits of geographical knowledge*. Minneapolis: University of Minnesota Press.

Rose, G. (2001). *Visual methodologies: An introduction to the interpretation of visual methods*. London: Sage.

Rose, G. and Tolia-Kelly, D. P. (2012). Visuality/materiality: Introducing a manifesto for practice. In: G. Rose, D. P. Tolia-Kelly, eds., *Visuality/materiality: Images, objects and practices*. Surrey: Ashgate, pp. 1–11.

Royal, C. (2008). Framing the internet: A comparison of gendered spaces. *Social Science Computer Review*, 26(2), pp. 152–169.

Rubinstein, T., Makov, S. and Sarel, A. (2013). Don't bi-negative: Reduction of negative attitudes toward bisexuals by blurring the gender dichotomy. *Journal of Bisexuality*, 13(3), pp. 356–373.

Schaeffer-Grabiel, F. (2004). Cyberbrides and global imaginaries: Mexican women's turn from the national to the foreign. *Space and Culture*, 7(1), pp. 33–48.

Seamon, D. (1979). *A geography of the lifeworld. Movement, rest and encounter.* London: Croom Helm.

Sedgwick, E. K. (1990). *Epistemology of the closet.* Berkeley: University of California Press.

Sedgwick, E. K. (1993). *Tendencies.* Durham: Duke University Press.

SedgwickE. K. (1999). *A Dialogue on love.* Boston: Beacon Press.

Sedgwick, E. K. (2003) *Touching feeling: Affect, pedagogy, performativity*, Durham: Duke University Press.

Senft, T. M. (2008). *Camgirls: Celebrity and community in the age of social networks.* New York: Peter Lang.

Shapiro, E. (2010). *Gender circuits: Bodies and identities in a technological age.* New York: Routledge.

Sharp, J. (2009). Geography and gender: What belongs to feminist geography? Emotion, power and change. *Progress in Human Geography*, 33(1), pp. 74–80.

Sharp, J. and Dowler, L. (2011). Framing the field. In: V. J. Del Casino, M. E. Thomas, P. Cloke and R. Panelli, eds., *A companion to social geography.* Oxford: Blackwell, pp. 146–160.

Sharp, J. P., Routledge, P., Philo, C. and Paddison, R. (2000). Entanglements of power: Geographies of domination/resistance. In: J. P. Sharp, P. Routledge, C. Philo and R. Paddison, eds., *Entanglements of power: Geographies of domination/resistance.* London: Routledge, pp. 1–42.

Shields, R. (2003). *The virtual.* London: Routledge.

Shields, R. (ed.) (1996). *Cultures of internet: Virtual spaces, real histories, living bodies.* London: Sage.

Siibak, A. and Tamme, V. (2013). Who introduced Granny to Facebook?: An exploration of everyday family interactions in web-based communication environments. *Northern lights: Film and media studies yearbook*, 11(1), pp. 71–89.

Silverstone, R., Hirsch, E. and Morley, D. (1992). Information and communication technologies and the moral economy of the household. In: R. Silverstone and E. Hirsch, eds., *Consuming Technologies.* London: Routledge, pp. 15–31.

Simpson, P. (2012). Apprehending everyday rhythms: Rhythmananalyis, time-lapse photography and the space-times of street performance. *Cultural Geographies*, 19(4), pp. 423–445.

Skelton, T. and Aitken, S. (eds.) (2016). *Establishing geographies of children and young people.* Singapore: Springer-Verlag.

Skelton, T. and Valentine, G. (eds.) (1998). *Cool places: Geographies of youth cultures.* New York: Routledge.

Skype.com (2014). About Skype – What is Skype? [Online]. Available at: http://www. skype.com/en/about/ [Accessed 2 May 2014].

Skype.com (2015). Moments of the month. [Online]. Available at: http://blogs.skype. com/2015/11/04/skype-moments-of-the-month-october-2015/ [Accessed 21 February 2016].

Slocum, R. (2008). Thinking race through corporeal feminist theory: Divisions and intimacies at the Minneapolis farmers' market. *Social and Cultural Geography*, 9(8), pp. 849–869.

Smith, J. (2014). Is it on pay-per-pew? Funerals to be live-streamed online as undertakers move their services into the digital age, *MailOnline*, 27 December 2014. [Online]. Available at: http://www.dailymail.co.uk/news/article-2888312/Is-pay-p ew-Funerals-live-streamed-online-undertakers-services-digital-age.html [Accessed 19 January 2015].

Spoonley, P. (1993). *Racism and ethnicity*. Auckland: Oxford University Press.

Stage, C. (2012). Screens of intensification: On DIY concert videos of Lady Gaga and the use of media interfaces as tools of experience intensification. *Journal of Aesthetics and Culture*, 4, pp. 1–11.

Statistic Brain Research Institute (2015). Skype company statistics. [Online]. Available at: http://www.statisticbrain.com/skype-statistics/ [Accessed 21 February 2016].

Stoker Bruenig, E. (2014). Figuring out how to mourn in the age of Skype, July 10, 2014. [Online]. Available at: http://www.theatlantic.com/national/archive/2014/07/fig uring-out-how-to-mourn-in-the-age-of-skype/374044/ [Accessed 20 January 2016].

Stone, A. R. (1991). Will the real body please stand up? Boundary stories about virtual cultures. In: M. Benedikt, ed., *Cyberspace: First steps*. Cambridge: The MIT Press, pp. 81–118.

StyleBistro (2013). How to look cute on Skype – Celebrity makeup artist Pati Dubroff's tips. [Online] Available at: http://www.stylebistro.com/Beauty+News/articles/7in THTSIsSq/ How+Look+Cute+Skype+Celebrity+Makeup+Artist [Accessed 3 May 2013].

Sullivan, N. (2003). *A critical introduction to queer theory*. Armidale: Circa, pp. 43–44.

Tarrant, A. (2010). Constructing a social geography of grandparenthood: A new focus for intergenerationality. *Area*, 42(2) pp. 190–197.

Tarrant, A. (2013). Grandfathering as spatio-temporal practice: Conceptualizing performances of ageing masculinities in contemporary familial carescapes. *Social and Cultural Geography*, 14(2), pp. 192–210.

Tarrant, A. (2014). (Grand)paternal care practices and affective intergenerational encounters using Information Communication Technologies. In: R. Vanderbeck and N. Worth, eds., *Intergenerational space*. New York: Routledge.

Tarrant, A. and Watts, J. (2014). Grandfathering and the embodiment of ageing masculinities. In: A. Tarrant and J. Watts, eds., *Studies of ageing masculinities: Still in their infancy?Representation of older people in ageing research series, 14*. London: Open University and Centre of Policy on Ageing.

Taylor, J. (1997). The emerging geographies of virtual worlds. *The Geographical Review*, 87(2), pp. 172–192.

Teather, E. (ed.) (1999). *Embodied geographies: Spaces, bodies and rites of passage*. London: Routledge.

The Skype Team (2015). Skype moments of the month September 2015. [Online]. Skype Blog. Available at: http://blogs.skype.com/2015/09/29/skype-moments-of-the-month-september-2015/ [Accessed 10 January 2016].

Thien, D. (2005). After or beyond feeling? A consideration of affect and emotion in geography. *Area*, 37(4), pp. 450–456.

Thien, D. (2011). Emotional life. In: V. Del Casino, M. E. Thomas, P. Cloke and R. Panelli, eds., *A companion to social geography*. Chichester: Blackwell, pp. 309–325.

Thompson, L. and Cupples, J. (2008). Seen and not heard: Text messaging and digital sociality. *Social and Cultural Geography*, 9(1), pp. 95–108.

Thrift, N. (2003). Closer to the machine: Intelligent environments, new forms of possession and the rise of the supertoy. *Cultural Geographies*, 10(4), pp. 389–407.

Thrift, N. (2004). Intensities of feeling: Towards a spatial politics of affect. *Geografiska Annaler: Series B, Human Geography*, 86(1), pp. 57–78.

Thrift, N. (2008). Non-representational theory: Space, place, affect. Abingdon: Routledge.

Thrift, N. and French, S. (2002). The automatic production of space. *Transactions of the Institute of British Geographers*, 27(3), pp. 309–335.

Todd, C. J. (2015). Sex and gender in World of Warcraft: Identities, love, and power. PhD thesis. University of Waikato. [Available at: http://researchcommons.waikato.ac.nz/handle/10289/9444].

Tolia-Kelly, D. (2006). Affect – an ethnocentric encounter? Exploring the 'universalist' imperative of emotion/affectual geographies. *Area*, 38(2), pp. 213–217.

Tuan, Y.-F. (1974). *Topophilia. Englewood cliffs*. New Jersey: Prentice-Hall.

Tucker, A. (2009). *Queer visibilities: space, identity and interaction in Cape Town*. Chichester: Wiley-Blackwell.

Turkle, S. (2011). *Alone together*. New York: Basic Books.

Underwood, H. and Findlay, B. (2004). Internet relationships and their impact on primary relationships. *Behaviour Change*, 21(2), pp. 127–140.

Valentine, G. (2006). Globalizing intimacy: The role of information and communication technologies in maintaining and creating relationships. *Women's Studies Quarterly*, 34(1/2), pp. 365–393.

Valentine, G. and Holloway, S. (2001a). A window on the wider world?: Rural children's use of information communication technologies. *Journal of Rural Studies*, 17(4), pp. 83–94.

Valentine, G. and Holloway, S. (2001b). On-line dangers?: Geographies of parents' fears for children's safety in cyberspace. *The Professional Geographer*, 53(1), pp. 71–83.

Valentine, G. and Holloway, S. (2001c). Technophobia: Parents' and children's fears about information and communication technologies and the transformation of culture. In: E. Hutchby and J. Moran-Ellis, eds., *Children, technology and culture: The impact of technologies in children's everyday lives*. London: Routledge, pp. 58–78.

Valentine, G. and Holloway, S. (2002). Cyberkids?: Exploring children's identities and social networks in on-line and off-line worlds. *Annals of the Association of American Geographers*, 92(2), pp. 302–319.

Valentine, G. and Skelton, T. (2008). Changing spaces: The role of the Internet in shaping deaf geographies. *Social and Cultural Geography*, 9(5), pp. 469–485.

Valkyrie, Z. C. (2011). Cybersexuality in MMORPGs: Virtual sexual revolution untapped. *Men and Masculinities*, 14(1), pp. 76–96.

Vancea, M. and Olivera, N. (2013). E-migrant women in Catalonia: Mobile phone use and maintenance of family relationships. *Gender, Technology and Development*, 17(2), pp. 179–203.

Vanderbeck, R.M. (2007). Intergenerational geographies: Age relations, segregation and re-engagements. *Geography Compass*, 1(2), pp. 200–221.

Vanderbeck, R.M. and Worth, N. (eds.) (2015). *Intergenerational space*. London: Routledge.

Waitt, G. R. and Duffy, M. (2010). Listening and tourism studies. *Annals of Tourism Research*, 37(2), pp. 457–477.

Wajcman, J. (2004). *Technofeminism*. Cambridge: Polity Press.

Wajcman, J. (2007). From women and technology to gendered technoscience. *Information Communication and Society*, 10(3), pp. 287–298.

White, M. (2015). *Producing women: The internet, traditional femininity, queerness, and creativity.* New York: Routledge.

Whitty, M. T. (2003). Pushing the wrong buttons: Men's and women's attitudes toward online and offline infidelity. *CyberPsychology and Behavior*, 6(6), pp. 569–579.

Whitty, M. T. (2005). The realness of cybercheating men's and women's representations of unfaithful internet relationships. *Social Science Computer Review*, 23(1), pp. 57–67.

Whitty, M. T. and Carr, A. N. (2005). Taking the good with the bad: Applying Klein's work to further our understandings of cyber-cheating. *Journal of Couple and Relationship Therapy* ,4(2–3), pp. 103–115.

Whitty, M. T. and Carr, A. N. (2006). *Cyberspace romance: The psychology of online relationships.* New York: Palgrave Macmillan.

Whitty, M. T. and Joinson, A. N. (2009). *Truth, lies and trust on the Internet.* London: Routledge.

Wikipedia (2015a). Skype. [Online]. Available at: http://en.wikipedia.org/wiki/Skype [Accessed 26 March 2013].

Wikipedia (2015b). Kitchen table: Women of color press 2015. [Online]. Available at: https://en.wikipedia.org/wiki/Kitchen_Table:_Women_of_Color_Press [Accessed 19 January 2016].

Wikipedia (2016). *Sara Ahmed.* [Online]. Available at: https://en.wikipedia.org/wiki/Sara_Ahmed#Life [Accessed 11 January 2016].

Wildermuth, S. M. and Vogl-Bauer, S. (2007). We met on the net: Exploring the perceptions of online romantic relationship participants. *Southern Communication Journal*, 72(3), pp. 211–227.

Wilken, R. and Goggin, G. (eds.) (2012). *Mobile technology and place.* London: Routledge.

Woleslagle, B. (2007). Halflings and ogres and elves, oh my! Sex, love, and relationships in EverQuest. In G. Herdt and C. Howe, eds., *21st century sexualities: Contemporary issues in health, education, and rights.* New York: Routledge, pp. 60–62.

Woods, J. (2008). Avatars and second life adultery: A tale of online cheating and real-world heartbreak. [Online]. *The Telegraph.* Available at: http://www.telegraph.co.uk/technology/3457828/Avatars-and-Second-Life-adultery-A-tale-of-online-cheating-and-real-world-heartbreak.html [Accessed 20 February 2010].

Wren, E. (2012). The biggest phobia in the world? 'Nomophobia' – The fear of being without your mobile – Affects 66 per cent of us, 8 May, 2012. [Online]. *MailOnline.* Available at: http://www.dailymail.co.uk/sciencetech/article-2141169/The-biggest-phobia-world-Nomophobia–fear-mobile–affects-66-cent-us.html [Accessed 21 January 2016].

Wright, M. (2010). Geography and gender: feminism and a feeling of justice. *Progress in Human Geography*, 34(6), pp. 818–827.

Wu, W., Fore, S., Wang, X. and Ho, P.S.Y. (2007). Beyond virtual carnival and masquerade in-game marriage on the Chinese internet. *Games and Culture*, 2(1), pp. 59–89.

Wykes, J. (2014). Introduction: Why queering fat embodiment? In: C. Pausé, J. Wykes, and S. Murray, eds., *Queering fat embodiment.* Farnham: Ashgate, pp. 1–12.

Yee, N. (2003). The Daedalus project: An ethnography of MMORPG Weddings. [Online]. *Nickyee.com.* Available at: http://www.nickyee.com/daedalus/archives/print/000467.php [Accessed 27 February 2013].

Yee, N. (2014). *The proteus paradox. How online games and virtual worlds change us – And how they don't*. New Haven: Yale University Press.

Yeh, S.J. and Sing, K.L. (2004). Living alone, social support and feeling lonely among the elderly. *Social Behaviour and Personality: An International Journal*, 32(2), pp. 129–138.

Young, I. M. (1990). *Throwing like a girl and other essays in feminist philosophy and social thought*. Indianapolis: Indiana University Press.

Index